THE FUTURE IS IN THE STARS

A Comprehensive Study of Biblical Astronomy

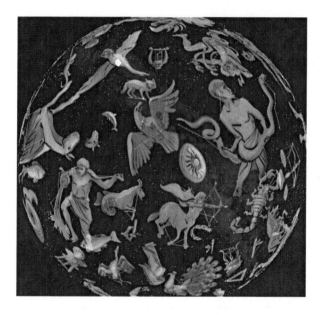

CLEVELAND CARTER

WestBow
PRESS

Copyright © 2012 by Cleveland Carter.

All rights reserved. No part of this book may be used or reproduced by any means, graphic, electronic, or mechanical, including photocopying, recording, taping or by any information storage retrieval system without the written permission of the publisher except in the case of brief quotations embodied in critical articles and reviews.

WestBow Press books may be ordered through booksellers or by contacting:

WestBow Press
A Division of Thomas Nelson
1663 Liberty Drive
Bloomington, IN 47403
www.westbowpress.com
1-(866) 928-1240

Because of the dynamic nature of the Internet, any web addresses or links contained in this book may have changed since publication and may no longer be valid. The views expressed in this work are solely those of the author and do not necessarily reflect the views of the publisher, and the publisher hereby disclaims any responsibility for them.

Any people depicted in stock imagery provided by Thinkstock are models, and such images are being used for illustrative purposes only.

Certain stock imagery © Thinkstock.

ISBN: 978-1-4497-5758-8 (hc)
ISBN: 978-1-4497-5755-7 (sc)
ISBN: 978-1-4497-5754-0 (e)

Library of Congress Control Number: 2012911383

Printed in the United States of America

WestBow Press rev. date: 07/03/2012

Contents

Glossary vii

Introduction xi

Chapter 1
The Origin of the Science of Astronomy 1

Chapter 2
Origin of the Zodiac 9

Chapter 3
The Celestial Equator and the Celestial Sphere 15

Chapter 4
The Ecliptic 18

Chapter 5
Equinoxes and Solstices 21

Chapter 6
The Story in the Zodiac 22

Chapter 7
The Galactic Equator 79

Chapter 8
Precession 82

Chapter 9
Precession and History 87

Chapter 10
The Hebrew Cycle and History 108

Chapter 11
Earth's Magnetic Field Decay 110

Chapter 12
Hebrew Feasts and Their Relation to Church Events 111

Chapter 13
The Great Pyramid of Giza.. 118

Chapter 14
The Bottom Line 125

Post Script 127

Bibliography 129

Glossary

Antediluvian: Those people who lived before the Genesis flood.

Biblical astronomy: Astronomy based on the biblical names and meanings of planets, stars, and constellations.

Booths (Tabernacles): A Jewish feast representing the gathering of the harvest. In Christendom, it represents Yeshua with us.

Celestial equator: The imaginary extension of earth's equator out to the end of space.

Celestial sphere: The dome of the sky surrounding earth.

Conjunction: The close proximity of the planets or moon with other stars.

Constellation: The formation of stars representing a figure with a meaning.

Ecliptic: The track of the sun through the sky.

Equinox: The point on the ecliptic where the celestial equator crosses it in the spring and autumn.

Firstfruits: A Jewish feast commemorating the first harvest of the Jews in the Promised Land after their exodus from Egypt. In Christendom, it represents Yeshua's resurrection.

Galactic equator: The imaginary line dividing the Milky Way into north and south.

Galactic plane: The imaginary line dividing the Milky Way into two symmetrical sections horizontally.

Galactic year: The time it takes for our solar system to complete one cycle around the galaxy's center. The galactic year would be equal to approximately 230 million earth years.

Grand year: Equal to the solar year.

Great year: The time it takes for the sun to make one complete cycle around the ecliptic. The great year is equal to about 25,700 earth years.

Lunar year: The time it takes for the moon to go through twelve full cycles. The lunar year is equal to 154.37 days.

Meridian: The imaginary line directly overhead that divides the sky into east and west.

Occult(ed): When one celestial object passes in front of another celestial object (from our vantage point on earth).

Passover: A Jewish feast commemorating the sacrifice of the lamb by the Jews preceding their exodus from Egypt. In Christendom, it represents Yeshua's sacrifice.

Pentecost: A Jewish feast commemorating the giving of the law. In Christendom, it commemorates the giving of the Holy Spirit.

Planisphere: A representation of the forty-eight original constellations.

Precession: The circular *wobble* of the earth's axis in space as the earth revolves around the sun. A period estimated at about 25,800 to 25,900 years.

Rosh Hashanah (Trumpets): A Jewish feast representing the Jewish new year. In Christendom, it represents His second coming foretold (announced) by the sounding of the trumpet.

Solstice: The time of year when the ecliptic and celestial equator are the farthest from each other (summer and winter).

Solar year: The time it takes for the sun to travel 360 degrees along the ecliptic in relation to the *fixed* stars of the zodiac.

Tropical year: The time it takes for the sun to travel along the ecliptic in relation to the four seasons.

Unleavened bread: A Jewish feast commemorating the exodus of the Jews from slavery in Egypt. In Christendom, it represents Yeshua, the sinless one.

Yom Kippur (Day of Atonement): A Jewish feast representing the removal of the nation's sin once every year. In Christendom, it represents Yeshua's second coming.

Zodiac: Technically, it is the twelve constellations on the ecliptic.

Introduction

From the beginning of time, men have looked up at the night sky and marveled at the beauty of the countless stars, a myriad of twinkling lights beckoning to the observer to contemplate the vastness of their habitation.

They have pondered the mysteries of the moon as it appears from nothingness, a thin white sliver cradling a ghost of a sphere. They have watched as it grew night by night in brightness, as it grew in fullness, until it became a ball of light illuminating the night sky. Its light has inspired poets and musicians to compose works unparalleled by any other inspiration except love. They have had questions about those celestial bodies: Where did you come from? Where are you going? What are you made of? How long will you live? What do you have to do with me, or what do I have to do with you? Is anyone out there?

They have, in the longings of their hearts, desired to unlock the deepest secrets hidden in the vast reaches of space. They have, in the imaginations of their minds, traveled to those distant stars—some blue, some white, some red, and some orange. They have traveled to the super giants and the dwarfs. They have traveled to where two, three, or even six stars, bound together by gravity, circle and interact with each other like a well-choreographed dance, which is so unlike that of our own solitary star. And they have uncovered deep secrets and brought them back to make this world, our home in space, a better place to live.

Does this sound like a pipe dream? Is it possible that somewhere, sometime in the future, this will really happen? We have been to the moon. Will man discover a method to get us beyond our solar system and back in less than a lifetime? Will man be around long enough to get to the stars? With the advent of the telescope—whether it is optical, radio, or otherwise—we seem to have drawn closer to the stars, and our knowledge has grown greatly.

Is there a future in the stars, or do the stars tell the future? There are those who believe that the stars and the planets and the sun and the moon rule their lives with a fate that they cannot control. They believe that their future is dictated by the position of the planets among the stars. Is Jupiter in Gemini?

Is Venus in Leo? Is Mars in opposition to Jupiter? Is Mercury in conjunction with Jupiter and the moon? The stars do tell the future; the stars also tell the past. I'm not talking about astrology, as it is known today. I'm talking about true astronomy. I'm speaking of the true study of the stars: where they are, their names, their relation to each other, and where the sun is in relation to them. I'm talking about solar and lunar eclipses and meteor showers. I'm speaking of discovering which star was the North Star 4,000 years ago. I'm speaking of discovering why the constellations are in the order they are and why the sun seems to travel in the direction it does.

The position and alignment of stars and planets simply cannot have any influence or persuasion on your life. Only you and God can do that. But the position of our sun and moon and the planets in the constellations at certain times through the coming ages, as revealed by computer models, can tell us much about the future, as a look at the past will show us.

This study is called biblical astronomy. Biblical astronomy is not new, but much of our information is. We are going to look at the zodiac and what it really means. Is it the story of cultural gods and mythical creatures battling for supremacy, or is it a different story? How many constellations are there? Have they changed since antiquity? Are there any modern inventions?

I say here that there are modern constellation *inventions*. We'll look at those modern constellations and their boundaries: are they accurate? We'll look at the celestial equator, an extension of the earth's equator out into space and why that is significant. We'll look at the *track* that the sun travels through the zodiac, called the ecliptic. We'll talk about the seasons and how they fit into the overall scheme of time and historical events. We'll look at which constellation the sun, moon, and planets were in when certain events took place. We'll examine the degrading magnetic field of the earth, and we'll look at some Jewish feasts and how they fit in with biblical astronomy.

There is more astronomy in the Bible and more Bible in astronomy than most people realize. There is more to the stars, the sun, the moon, and the planets than just pretty pictures from NASA; there is more than $E=MC^2$. The physics of astronomy is fascinating, and the pictures are awesome, but the rest of the story will make you think more than twice about your future.

CHAPTER 1

The Origin of the Science of Astronomy

Although the origin of the constellations is somewhat obscure, in the traditions of the nations where the constellations are preserved and observed there are no doubts about the origin of the science of astronomy. It has always and everywhere been traced back to the earliest race of man, to those who lived before the flood.

Claudius Ptolemy (ca. AD 90–ca. 168) was a Roman citizen of Egypt. He was a mathematician and astronomer. Ptolemy was the author of several scientific treatises. The first is the astronomical treatise now known as the *Almagest*. The *Almagest* is his only surviving comprehensive ancient treatise on astronomy. Ptolemy's catalogue was based almost entirely on an earlier one by Hipparchus.

Hipparchus (ca. 190–ca. 120 BC) is considered the greatest ancient astronomical observer and, by some, the greatest overall astronomer of antiquity. He was the first astronomer whose quantitative and accurate models for the motion of the sun and moon survive. It is commonly believed and accepted that he made use of the observations and perhaps the mathematical techniques accumulated over centuries by the Chaldeans from Babylonia. He completed his star catalogue in 129 BC.

Hipparchus is generally given credit for discovering precession, which will be explained in detail later, because he could not reconcile the position of stars in his time with those noted by earlier Babylonian astronomers. Even so, there is documentation to the contrary. (See the following note on Aristarchus.)

Aristarchus (310–ca. 230 BC) is the attributed author of two ancient manuscripts preserved by the Vatican library and containing estimates of the length of the year. These manuscripts have two different values for the length of the year. The differing numbers represent the sidereal year and the tropical year; each has a distinct time period. (The sidereal and tropical years will be explained later.) Aristarchus's work was based on the prior work of Meton of Athens. Both denominators can be related to Aristarchus, whose summer solstice was 152 years after Meton's. The time difference between the sidereal year and the tropical year is identical to precession.

The learned Jewish historian Josephus refers to ancient writers as his authorities; we still have the names of these writers, but none of their writings. He attributes the invention of the science of astronomy to the family of Seth, the son of Adam. Because men then had life spans of nearly a thousand years, they were personally able to see from actual observation how the celestial bodies returned to the same positions in set cycles and periods. In later ages, it has required the labor of successive generations to verify those observations. Josephus mentions a "pillar of stone" in Siriad (present-day Egypt).

> [T]hey also were the inventors of that peculiar sort of wisdom which is concerned with the heavenly bodies, and their order. And that their inventions might not be lost before they were sufficiently known, upon Adam's prediction that the world was to be destroyed at one time by the force of fire, and at another time by the violence and quantity of water, they made two pillars, the one of brick, the other of stone: they inscribed their discoveries on them both, that in case the pillar of brick should be destroyed by the flood, the pillar of stone might remain, and exhibit those discoveries to mankind; and also inform them that there was another pillar of brick erected by them. Now this remains in the land of Siriad to this day.[1]

I will say more about this "pillar of brick" later.

Giovanni Battista Riccioli (1598–1671) was an Italian astronomer who thought the ancient Arab names of stars and constellations were antediluvian.

Ja'far ibn Muhammad Abu Ma'shar al-Balkhi (787–886), also known as Albumasar, was a Persian mathematician, astronomer, and philosopher. Albumasar credits the Persian zodiac to the first Hermes, the second Hermes, and Ascalius. Hermes, meaning "great," and Ascalius, meaning "the skilful," are believed to be only epithets. On ancient Arab authority, "the first Hermes," to whom ancient writers so often credit the origin of human science, refers to Enoch. It is presently not known to whom Ascalius refers.

Manthos Oikonomou (1754–1822), a Greek scholar, is referred to as having said that the first Hermes was antediluvian.

El Macinus, Abulfaragius and other Arabic writers from the early thirteenth century call Enoch "Edris," or "the Glorious." They say he was skilled in astronomy and other sciences and that he was the same as Hermes. The Jews (as in the Targum of Jonathan) called him "the great scribe" and said that he was the first who composed books of astronomy.

Jean-Étienne Montucla (1725–1799), a French mathematician, refers to John Malalas (ca. 491–578), a Byzantine chronicler from Antioch, as having said that Seth himself divided the sky into constellations.

Sir W. Drummond wrote in 1824 that he believed that at some remote period there were mathematicians and astronomers who knew that the sun was at the center of our solar system and that the earth was itself a planet that revolved around it. He also said that Origen said that it was asserted in the *Book of Enoch [I]* that in the time of that patriarch, the constellations were already named and divided.

Origen (ca. AD 185–254) is thought (not only by Drummond but also by others) to allude to a *Book of Enoch [I]*, which, to our knowledge, no longer exists. We do, however, have another *Book of Enoch [II]*, which describes the number of constellations and other astronomical phenomena.

The earliest known star catalogues were compiled by the ancient Babylonians of Mesopotamia in the late second millennium BC, during the Kassite Period (ca. 1531–ca. 1155 BC). They are best known by their Assyrian-era name: *Three Stars Each*. These star catalogues were astronomical symbols written on clay tablets.

The Egyptians, on whose early monuments the zodiac signs are found, acknowledged that they derived their astronomy from the Chaldeans. The Chaldeans attributed their science of astronomy to Oannes, who is known to us as Noah. The Chaldeans have been traced back to Arphaxad, Noah's grandson. The Hebrews, Persians, and Arabs traditionally imputed its invention to Adam, Seth, and Enoch.

Soon after the usual date assigned to Noah's flood, astronomy is found highly cultivated in the empire of China. Equally early records of observed eclipses were preserved at Babylon, proving the science of astronomy was attained there.

Modern writers who acknowledge the authority of the Hebrew Scriptures, if only as historical texts, generally attribute astronomy's origination and transmission to the ancient nations to the antediluvian patriarch, Noah. Those who do not admit that authority claim the age astronomy, from its internal evidence, to be between five thousand and six thousand years. (Is that any surprise, since the dating of the antediluvian patriarchs is the same?)

There is no evidence that the science of astronomy ever existed separate from the constellations; since they are essential to astronomy, their constant connection with it can only be explained by their having been set forth at the same time and by the same people.

Concerning the *Book of Enoch [II]*, some have doubted the reliability of this writing on the basis that Enoch's writing is a combination of spiritual and scientific material and was not accepted as canonical (biblical) Scripture by the Christian church. Here we are not concerning ourselves with the spiritual debate but with the scientific aspect of his writing. Biblical writer Jude, in his letter to the Christians of his time, references a prophecy of Enoch in Enoch's writing, which proves that this writing was recognized as legitimate in the first century AD.

Jude 14–15 states,
> Now Enoch, who lived seven generations after Adam, prophesied about these people. He said, "Look, the Lord is coming with thousands of his holy ones. He will bring the people of the world to judgment. He will convict the ungodly of all the evil things they have done in rebellion and of all the insults that godless sinners have spoken against him." (NLT).

Having been given the perspectives of authors and civilizations from various countries, cultures, and eras on the origins of the science of astronomy, it seems that the only evidence we have as to its origin are the extant records of ancient tradition that attribute it to the family of Seth. Therefore, since no conflicting evidence exists, should we not accept this information as conclusive? Why should we not credit Seth, who seems to have calculated the movements of the sun, moon, stars, and planets so precisely, with being the father of astronomy?

Though tradition has it that Seth instructed his offspring who instructed their offspring down to Enoch in the science of astronomy, Enoch seems to be the first to have written about it.

The following are excerpts from the *Book of Enoch*. Note that all references to the *Book of Enoch* from this point on refer to Enoch's second book, as translated by Schodde.

Enoch 37:1
> The second vision of wisdom which Enoch, the son of Jared, the son of Mahalaleel, the son of Cainan, the son of Enos, the son of Seth, the son of Adam, saw . . . [2]

Notes on Enoch 37:1
Enoch is identifying himself by his ancestry. What is interesting here is that all of these people, back to and including his great, great, great, great grandfather, Adam, were still living. You can do the math from Genesis to prove this.

Enoch 72:4–9
> 4. And first comes forth the great luminary called the sun; and his circuit is like the circuit of the heavens, and he is entirely filled with flaming and heating fire.
> 5. The wagons on which he ascends are driven by the wind, and the sun descending disappears from the heavens and returns through the north in order to reach the east, and is led that he comes to that portal and shines on the surface of heaven.
> 6. And thus he comes forth, in the first month, in the great portal, and he comes forth from the fourth of these six portals toward the east.

> 7. And in that fourth portal, from which the sun comes forth in the first month, there are twelve window openings, from which a flame proceeds when they are opened in their time.
> 8. When the sun rises from the heavens he comes out of that fourth portal thirty mornings, and descends directly into the fourth western portal of heaven.
> 9. And in those days the day is daily lengthened, and the nights nightly shortened to the thirtieth morning.[3]

Notes on Enoch 72:4–9

4–5: Enoch here simply describes the sun and its "circuit" from rising to setting to rising again.
6–9: He here uses the term "in the first month" to describe the first month of the year. The term "portal" is a constellation in the zodiac. Then Enoch adds that the sun comes from the fourth portal. Verse 9 clarifies this.
9: Enoch is saying that the days are getting longer and the nights shorter. This phenomenon occurs at the spring equinox. And so, Enoch is describing the time of the year, the beginning of spring.

Enoch 73:1, 74: 10–12

> 1. And after this law I saw another law with reference to the smaller luminary whose name is moon.
> 10. And in those days, if five years are taken together, the sun has thirty superabundant days; and all the days which belong to him for one of these five years, when they are full, are three hundred and sixty-four days. 11. And the superabundance of the sun and of the stars is six days; of five years, each at six, are thirty days, and the moon recedes from the sun and the stars thirty days.
> 12. And the moon brings in all the years exact, so that their place neither precedes nor recedes ONE day, but she changes the years with exact justice in three hundred and sixty-four days.[4]

Notes on Enoch 73:1, 74:10–12

Here Enoch is describing the lunar year. It may sound a little complicated in his terminology, but all he is stating is that the lunar year is 364 days long.

THE FUTURE IS IN THE STARS

Enoch 75:1–3
> 1. And the leaders of the heads of the thousands, who are over all creation and over all the stars, are also with the four intercalary days, which cannot be separated from their places, according to the whole reckoning of the years, and these serve the four days which are not counted in the reckoning of the years.
> 2. And on their account men make a mistake in them, for these luminaries serve in reality on the stations of the world, one in the first portal and one in the third portal and one in the fourth portal and one in the sixth portal; and the harmony of the course of the world is brought about by its separate three hundred and sixty-four stations.
> 3. For the signs and the times and the years and the days, these the angel showed to me, he whom the eternal Lord of glory had placed over all the luminaries of heaven in the heavens and in the world, that they should rule on the surface of the heavens, and be seen on the earth, and be leaders for the day and for the night, viz. the sun and the moon and the stars and all the serving *creatures* who keep their course in all the chariots of heaven.[5]

Notes on Enoch 75:1–3
Here Enoch is distinguishing between the solar year and the lunar year. The lunar year is about four days shorter than the solar year. Enoch's method of distinguishing time then was as accurate as ours is today using computers, but somewhere along the line this reckoning of time was lost. We now observe time (months and years) by the Gregorian calendar. It was introduced by Pope Gregory XIII, after whom the calendar was named, by a decree signed on February 24, 1582. The reformed calendar was adopted later that year by a handful of countries, with other countries adopting it over the following centuries.

The motivation for the Gregorian reform was that the Julian calendar (invented by Julius Caesar and used in Western countries until the Gregorian calendar), assumes that the time between vernal equinoxes is 365.25 days, when in fact it is about eleven minutes less. The error between these values was about ten days, which is exactly what Enoch said. Therefore, the Julian calendar accumulated time at the rate of about three days every four centuries, resulting in the equinox occurring on March 11 and moving steadily earlier in the calendar by the time of the reform. Since the equinox was rightfully tied to the celebration of Easter, the Roman Catholic church considered that this steady movement was undesirable, hence the Gregorian calendar.

1. Flavius Josephus. William Whiston, trans. *Antiquities of the Jews*. Project Gutenberg, 2009. http://www.gutenberg.org/. eBook.
2. Enoch. *Book of Enoch*. George H. Shodde trans. W. F. Draper, 1882, 52. http://www.pigspine.com
3. Ibid., 83.
4. Ibid., 85,86.
5. Ibid., 86.

CHAPTER 2

Origin of the Zodiac

What is the zodiac? Technically, it entails the twelve constellations on the ecliptic. But for our purposes, the zodiac entails the original forty-eight constellations.

Leaving the science of astronomy and turning to the constellations themselves and what they represent, I first must point out that Enoch, in his description of the heavens and the sun's path through it, lists twelve "portals" and "many windows" in the heavens. These are the constellations. It will become obvious that the constellations were not invented around bright stars. Although many of the constellations contain bright stars, not all do. In fact, when we consider the 168 brightest stars, not one of them is in the constellation Ara. Therefore, there must be another explanation for the constellations.

Here's an excerpt from the *Book of Enoch* with the description:

Enoch 72:3
>And I saw six portals, out of which the sun ascends, and six portals into which the sun descends; the moon also rises and sets in these portals, and the leaders of the stars and those led by them; six in the east and six in the west, and all, each after the other, aright; also many windows to the right and to the left of these portals.[1]

Again, although Enoch, in this book, does not specifically state how many of these "windows" there are, tradition has it that there were thirty-six "windows." Early zodiacs, such as the zodiac of Dendereh, dating back to 2000 BC or earlier, bear this out. It doesn't take a rocket scientist to do the math and see that the twelve portals and thirty-six windows add up to a total of forty-eight constellations. At present, according to the International Astronomical Union (IAU), there are eighty-eight constellations. And so the question is this: why is there a difference in the number of constellations? And which, if either, is correct?

Since the time of Enoch, various cultures have added constellations to the star maps. New constellations have been *invented* and added as late as the eighteenth century AD. Nicolas Louis de Lacaille (1713–1762) was a French astronomer. He is noted for his catalogue of nearly 10,000 southern stars. This catalogue, called *Coelum Australe Stelliferum*, was published posthumously in 1763. It introduced fourteen new constellations, which have since become standard.

For our purpose, dealing with the most ancient astronomy and the constellations known to Enoch and to Seth, I must emphasize that the original forty-eight constellations are the ones we are concerned with. Let me say here that I have no problem with some of the *new* constellations that have since been added, but they do not fit into the scheme of the picture I am about to paint. It is like putting an oak tree in a painting of an orange grove. I will have more to say on this later.

I am sure that most people know that the constellations represent certain figures. Why are these particular "symbols" or "figures" used? Could not others have marked out the division of the sun's path or that of the moon or planets just as well? Especially since very few of these "figures" actually look like the figures they supposedly represent. One suggestion is that the zodiac is representative of the seasons with three constellations for each of the four seasons. I have heard this as an explanation on numerous occasions. But it should be pointed out that, with the phenomena of the cycling year, they cannot be consistently made to agree with the changing climate everywhere on earth all of the time. How can we "force" the constellations to agree with the seasons at the time when the ice age covered so much of the earth with ice? Who was there in the middle of a glacier to start planting or harvesting crops because the stars said so? Can the mythology then of nations give rise to a reasonable explanation? Though reflecting their "shadows," varying so greatly from one nation and culture to the next, mythologies are insufficient to account for their imagery. But if their traditional names and figures can be shown to symbolize or represent biblical prophecy, revealed in the earliest age of man (the time of Seth and Enoch), then the use of these representations makes sense.

So, as tradition has it, Seth instructed his offspring and so on down to Enoch in not only the science of astronomy but also in the meaning of the zodiac.

However, there may be more than just tradition. The Dendereh planisphere depicts not only the zodiac (the central circle) but also the signs the sun was in at the equinoxes and solstices. It is believed that this represents the time that the planisphere would have been made. The reason for believing this is the fact that other planispheres from other times contain the same information. We can verify when these planispheres were made. The zodiacs are the same, but the signs the sun was in at the time of the equinoxes and solstices are different and so represent the different times when these planispheres were made.

If this is true, and I believe it is, the date of the Dendereh planisphere would have to be around 4000 BC.

As we will see later, this date will be extremely important.

Credit: Louvre Museum. US Public domain.

This then begs the question: how did Seth come to know the meaning? Did Seth actually "invent" the constellations and their meanings as tradition suggests? For this answer, we turn our attention to the Bible. Let me at this point say that, if in no other way, we can use the Bible as a historical record. I believe it to be more—much more—but I take it to also be an accurate, reliable source for historical purposes. I must confess though that, in context of our study, we must use the Bible in both historical and spiritual context.

In the following Scriptures, words in **boldface** are from my emphasis.

Genesis 1: 14–18
> [14]And God said, Let there be lights in the firmament of the heaven to divide the day from the night; and **let them be for signs, and for seasons,** and for days, and years: [15]And let them be for lights in the firmament of the heaven to give light upon the earth: and it was so. [16]And God made two great lights; the greater light to rule the day, and the lesser light to rule the night: **he made the stars also.** [17]And **God set them in the firmament of the heaven** to give light upon the Earth, [18]and to rule over the day and over the night, and to divide the light from the darkness: and God saw that it was good. [19]And the evening and the morning were the fourth day. (KJV paraphrased).

Notes on Genesis 1:14–18
God created the sun, moon, and stars and placed them in space for light, to tell the seasons and to be for signs.

Psalm 8:3
> When I look at your heavens, the work of your fingers, the moon and the stars that **you have established.** (NRSV).

Notes on Psalm 8:3
God made the stars and set them where they are. The word "established" means not only made but also placed where they are.

Job 38:31–33
> ³¹"Can you bind the cluster of the **Pleiades,** or loose the belt of **Orion**? ³²Can you bring out **Mazzaroth (zodiac constellations)** in its season? Or can you guide the **Great Bear** with its cubs? ³³Do you know the ordinances of the heavens? Can you set their dominion over the earth?" (ASV paraphrased).

Notes on Job 38:31–33
Job, considered the oldest book in the Bible, predating even Genesis and Exodus (the order of the books of the Bible is not chronological), names constellations and asterisms. And it even states astronomical properties: the Pleiades, as a star system, are gravitationally bound.

32: "The Great Bear" is a mistranslation. I'll explain that later when I talk about the constellation Ursa Major and Ursa Minor.

33: God asks Job, "Can you explain the physics of space?" and "Can you set all of the celestial bodies in orderly motion like I have?" God is telling Job that He is responsible for **all** that is out there.

Psalm 147:2–4
> ²The LORD builds up Jerusalem; he gathers the outcasts of Israel. ³He heals the brokenhearted, and binds up their wounds. ⁴He **determines the number** of the stars; he **gives to all of them their names.** (NRSV).

Notes on Psalm 147:2–4
2–4: God determined the number of stars and God named them. God made man (and woman). God gave Adam the job of naming the animals, plants, and everything else on earth. But God determined the number of stars He would put "out there," and He named the stars.

Psalm 19:1–4
> ¹The heavens tell of the glory of God. The skies display his marvelous craftsmanship. ²Day after day they continue to speak; **night after night they make him known. ³They speak without a sound or a word; their voice is silent in the skies; ⁴yet their *message* has gone out to all the earth, and their words to all the world.** (NLT).

Notes on Psalm 19:1–4
1: The heavens—the sun, moon, and stars—tell God's glory.
2–4: The term "day after day" references the sun; the term "night after night" specifically refers to the stars only; the moon is not referenced here because the moon is not visible "night after night." The stars speak in a silent voice to all the earth to make God known.

But how do the stars, the forty-eight constellations of the heavens, make Him known? They do indeed tell a story. It is not the stories of the Greek or Roman gods and creatures of their mythologies or those of any other individual nation or culture. They, in a sense, tell the story of the history of the world, the history of all of mankind. It is the story of the past, the present, and the future. They tell the story of the One who is at the center of the history of the world, not necessarily in time but the focus of events. They tell the story of the Christ, or the gospel story.

1. Enoch. *Book of Enoch*. George H. Shodde trans. W. F. Draper, 1882, 83. http://www.pigspine.com

CHAPTER 3

The Celestial Equator and the Celestial Sphere

Before we turn our attention to the forty-eight constellations and "read" the story, we must lay some scientific groundwork. This groundwork is the foundation for understanding how the constellations tell us the past, the present, and the future.

Do you know what is meant by the term *equator*? It's an imaginary line going around the earth at its widest point east to west. It is that imaginary line that separates the northern and southern hemispheres. Have you ever been any place on the earth at the equator? I have, in Ecuador, South America. I saw a monument that shows the dividing line between the northern and southern hemispheres, and there is a line in the concrete around the monument. Interestingly, if you stand directly over that line with half of your body on one side of the line and the other half of your body on the other side of the line, half of you is in summer and the other half of you is in winter (or spring and fall). But it doesn't feel any different.

Just as the earth's equator is not "real," that is, not something physical, imagine a line extending from the earth's equator out into space as far as it can go—to the "ends" of the universe. That would be what is called the celestial equator. The celestial equator is perpendicular (ninety degrees) to the earth's axis. How far does it go? Well, how big is the universe? Let's just say it's big (one estimate places it at about two-hundred-fifty billion light years) and the celestial equator extends all the way to its "end."

The following illustration shows the relation of the celestial equator to the earth as I have stated. Some of the other items in the drawing will be discussed later.

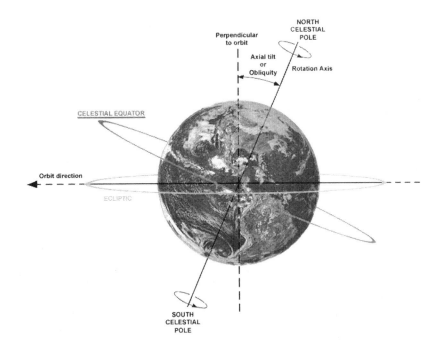

Credit: Dennis Nilsson, Creative Commons Attribution 3.0 Unported.

There are several theories about the "shape" of space. Most of those theories suppose space to be confined on a "bubble" or sphere in some fashion. The Bible supports this. Imagine space extending out around the earth, around our solar system, around our galaxy, and around all of the other galaxies out there. Picture all of the stars, galaxies, nebulae, and other celestial *stuff* out there scattered from here, in all directions, hundreds of billions of light years through space. That is the celestial sphere. And the celestial equator extends from the earth out through all of those things in that "bubble."

Can you remember, probably from some time in school, when you learned that the earth is tilted in space in relation to its orbit around the sun? That is the earth's spin axis. The earth's spin axis is tilted by 23.5 degrees with respect to the earth's orbital plane around the sun as shown below.

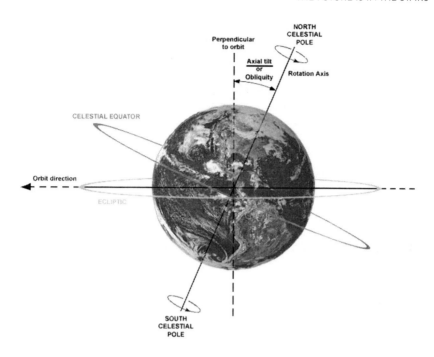

Credit: Dennis Nilsson, Creative Commons Attribution 3.0 Unported.

The earth's spin axis is not fixed in space; however, it wobbles a little, like a spinning top or the *free* end of a spinning gyroscope on a tabletop (and we'll talk more about that later. Let's just say that it is tilted and it is this 23.5 degrees tilt of the earth's axis which causes the seasons. Through the course of the year, the sun spends six months above the equator (celestial and terrestrial, about March 21 to about September 20) and six months below (about September 21 to about March 20). These dates can vary by a day.

Celestial objects near the celestial equator are visible worldwide. This is important because the majority of inhabitable land lies far enough north of the Antarctic region and far enough south of the Arctic region that the zodiac is visible from any hospitably inhabitable land. This means that the gospel story in the stars is seen from all those lands.

CHAPTER 4

The Ecliptic

Think again of the celestial sphere. As the earth orbits around the sun over the course of a year, we observe the sun to track out a full circle of 360 degrees around the celestial sphere. This translates into just under one degree per day.

This track of the sun on the celestial sphere is called the ecliptic. The word *ecliptic* is taken from the Greek word that means "the place where eclipses take place." As you can see in the diagram below, the ecliptic (the solid circle) serves as the center of the zodiac. Not only does the sun travel on the ecliptic, but the moon and the planets in our solar system are very nearly on it. They vary no more than three degrees from the ecliptic except for Mercury, which can be seven degrees, and their orbits cross the ecliptic at various places at various times.

Reproduction of Albumasar's Zodiac. Public domain.

This picture taken by a 1994 lunar spacecraft shows this very well. We see the moon with the sun rising over it and the planets Saturn, Mars, and Mercury to the left.

The Plane of the Ecliptic. NASA image. Public domain.

So now when you hear someone speaking of astrology, say that the planets are in alignment, you can tell him or her that you know the planets are always nearly in alignment (technically). They just aren't always in close visual proximity to one another (in conjunction). As we look further into the zodiac and the science of astronomy, we'll see more evidence that disproves modern-day astrology.

Again, the earth is tilted at a 23.5 degrees angle in space and the celestial equator is perpendicular to the earth's axis. The ecliptic is parallel to the earth's orbit around the sun. This means that the celestial equator is inclined 23.5 degrees to the ecliptic plane and intersects with it in two places (on opposite sides of the earth). Why is this important? The location of the sun on the ecliptic in relation to the celestial equator tells us which season we are in. Notice in the drawing below the intersection of the celestial equator and the ecliptic.

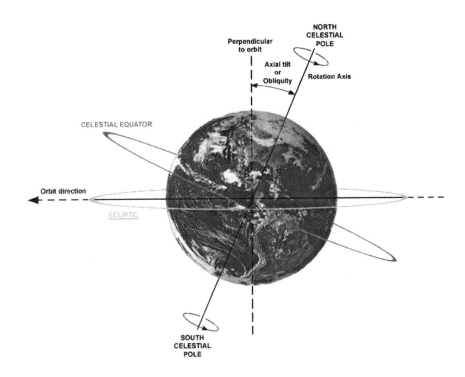

Credit: Dennis Nilsson, Creative Commons Attribution 3.0 Unported.

CHAPTER 5

Equinoxes and Solstices

A brief explanation of equinoxes and solstices is needed here because they play an important role in determining a timeline for historical events. The points where the celestial equator intersects the ecliptic are called the equinoxes (spring and fall). And the points where the celestial equator is furthest from the ecliptic are called the solstices (summer and winter). You may have heard these terms. The times for these occurrences are as follows:

March 20–21 for the spring equinox
June 20–21 for the summer solstice
September 22–23 for the fall equinox
December 21–22 for the winter solstice

Due to the way we now track time, that is, by our present calendar, the above dates will always hold true. The spring equinox, also known as the vernal equinox, has also been known by the term "First Point of Aries." This term was coined by ancient astrologers believing that the first sign of the zodiac is Aries. Astrology then lists the rest of the signs in clockwise motion. As to the direction to follow the signs of the zodiac for telling time, that is, projecting the future, which is what astrology purportedly does, astrology has it backward. The proper motion for projecting the future is counterclockwise. We know this to be the true motion because this is the direction God set the sun in along the ecliptic. It is further illustrated by the directions God gave to Moses for the priests in the tabernacle (and later in the temple) to perform their duties. They would enter the tabernacle (and later the temple), turn right, and move counterclockwise to each station. Also, when Israel's tribes encamped around the tabernacle in the wilderness, they set up their camps in a counterclockwise direction starting from the east (Numbers 2). God's orderliness shows up even in places we normally don't think about.

Now that we have a basic understanding of the celestial equator, the ecliptic and the equinoxes and solstices, we can move on to the story in the zodiac.

CHAPTER 6

The Story in the Zodiac

We now want to look at the story the zodiac tells. As I said earlier, the original forty-eight constellations tell the story of world history past, present, and future. It is the story of Christ.

Now, having forty-eight constellations to tell the story, where do we start? To begin with, we can narrow the choice by the fact that the sun, which is our main player in determining days, months, and years, "travels" along the ecliptic. Therefore, we can reduce our choices of a starting point down to twelve constellations. This is better than forty-eight, but we still need the right starting point.

Reproduction of Albumasar's Zodiac. Public domain.

Remember I said that astrology says Aries is the beginning. That may sound good since, as we shall see later, Aries (the ram) represents Christ, the sacrifice. But as with everything that Satan gets his hands into, astrology has perverted the truth. The starting point can be determined by going back to the beginning of world history. In the book of Genesis, after Satan had deceived Eve and Adam, God made a promise to them (in chapter 3). He said in verse 15, "From now on, you and the woman will be enemies, and your offspring and her offspring will be enemies. He will crush your head, and you will strike his heel." This is God's promise to send a world Savior through the offspring of the woman.

This makes the choice of a starting point easy, since only one constellation on the ecliptic represents the woman brining forth the promised offspring. That constellation is Virgo. Therefore, Virgo is the beginning of the zodiac. How else do we know this? The Sphinx tells us. Yes, the Sphinx.

Most references say that the Sphinx is a mythological creature depicted as a recumbent lion with a human head. In Egypt, some Sphinxes are the body of a lion with the head of a Pharaoh. It seems that the only constant is the lion's body. But, in fact, the Sphinx of Giza, which is believed to be the oldest sculpted figure, dated to ca. 2623 BC (but probably is older), has the body of a lion and a woman's head.

Credit: w:es:Usuario:Barcex, Creative Commons Attribution 3.0 Unported.

What is the significance of this? Since it faces directly east, it faces toward the exact position of the sun on the Vernal Equinox. And since the sun moves from east to west, this tells us which way the zodiac is to be followed. And so, the Sphinx also gives us the real starting place in the zodiac. The

Sphinx shows the beginning and the end. I do not believe the Sphinx to be "circumstantial evidence," but I will keep with the biblical evidence for the beginning. Therefore, we start with Virgo and end with Leo.

The story line moves in a clockwise motion. Another fact we have to keep in mind is the "windows" Enoch spoke of. As I said previously, these are the other thirty-six constellations. Each of the twelve constellations on the ecliptic has another three associated with it to complete its "chapter" of the story. These groupings of three constellations are called *decans*, a Latin word which means "house." And so these groupings of decans portray Christ and His story. The first grouping portrays the first coming; the second grouping portrays His work and the blessings He gives; the third grouping portrays His second coming and judgment of sin.

At this point, I must digress. I earlier said that modern astronomy shows eighty-eight constellations. I also said that I have no problem with the addition of *some* of these constellations. Well, that's not entirely true. Allow me to explain. I do have a problem with new constellations in the sense that when placed anywhere in the planisphere, even in "blank" spaces, they are adding extraneous information to the story because they have nothing to do with the true story, and so they detract from the true story, add confusion to the truth, and do not belong there.

One example of this is the "new" constellation Telescopium. It was invented by the French astronomer Nicolas Louis de Lacaille in the eighteenth century in honor of the science of astronomy. It is located south of Sagittarius and near Ara. Originally, de Lacaille "borrowed" stars from Sagittarius and Scorpio for his invention, but the International Astronomical Union placed those stars back where they belong so that Telescopium now does not impinge on any of the original constellations of the planisphere.

Other of these "new" constellations, mostly in the southern hemisphere, never did, and hopefully never will, impinge on any of the original constellations. But not all of these "new" constellations are so. Canes Venetici is one example. Located between Bootes and Ursa Major, the constellation depicts two hunting dogs, on leash, held by Bootes, a hunter chasing Ursa Major. Supposedly Ursa Major is the Great Bear. Only here, Ursa Major is not a bear but a donkey.

THE FUTURE IS IN THE STARS

Credit: Johannes Hevelius. Public domain.

What's wrong with this picture? Just to give you a taste of what's to come . . . First, the dogs do not belong there. Bootes is not a hunter. Ursa Major is not a donkey or a bear. So, as we go through the constellations, we will see more that is wrong. And because Bootes is in the first grouping of constellations, it won't be long.

I need to say here that just as the wording of a verse in the Bible has one interpretation of the words but can have many applications and therefore many "meanings," according to those applications, so it is with the signs of the planisphere. Each sign will have one interpretation but may have more than one meaning, depending on its application in time.

Now, let's start our story with Virgo and see what Virgo and its decans tell us.

The first grouping—Virgo, Libra, Scorpio, and Sagittarius—shows the Redeemer's first coming. Virgo depicts our Redeemer's "virgin" birth. Libra, the scales, tells how our Redeemer paid the required price (scales were used for weighing the price in trades). Scorpio shows a scorpion wounding a man's heel while being trodden under foot in fulfillment of Genesis 3:15: "The seed of the woman will crush your head, and you will strike his heel." (NLT paraphrased). Centaurus shows a dual-natured being (God-man) victorious over the serpent.

THE SIGN VIRGO AND ITS DECANS

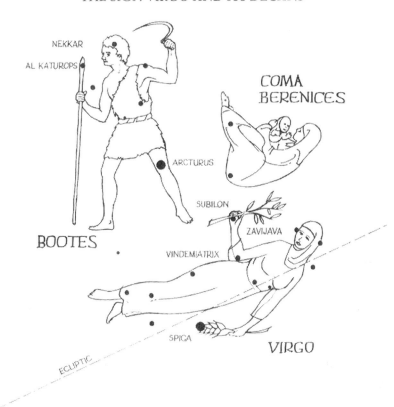

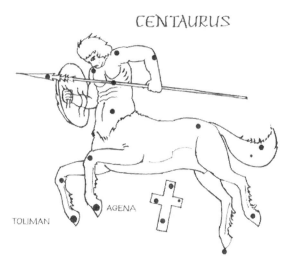

Reproduction from Albumasar's Zodiac. Public domain.

Virgo: The Seed of the Woman
>Isaiah 7:14
>Therefore the Lord Himself will give you a sign: Behold, the virgin shall conceive and bear a Son, and shall call His name Immanuel. (KJV).

Virgo and its decans tell of the virgin birth of the Desired Son that would be a despised sin offering to redeem man. Virgo means "a virgin." *Virga* in Latin means "a branch." Both words are used to refer to Jesus in the Latin Vulgate. Isaiah 11:1 refers to Jesus as "the Branch" and Matthew 1:23 refers to Jesus as virgin-born.

Virgo is a virgin woman holding an ear of wheat in her left hand and a branch in her right hand. The ear of wheat refers to the Seed of the woman who is to bruise the head of the serpent as was described in Genesis 3:15. In John 12:24, Jesus refers to himself as the "Kernel of Wheat" that would fall to the ground and die in order to bring forth fruit. This obviously is a reference to His work in His first coming (that is, His death and resurrection). The "fruit" He would bring forth refers to the believers. The virgin is actually a reference to the "first church," the Jews, and not Mary, the daughter of Heli. The Messiah had to come through the Jews. He is always associated with the "first church," the Jews.

Stars in Virgo and their meaning:
>Spica (located in the wheat ear, the seed carried by the virgin):
>The Branch (the physical manifestation of the Messiah)
>Tsemech: The Branch (the Hebrew word refers exclusively to the spiritual manifestation of the Messiah)
>Zavijaveh: The Gloriously Beautiful (the spiritual radiance)
>Vindemiatrix: The Branch Who Comes (the promised Messiah)
>Subilah: Who Carries (the spiritual sustainer)
>Subilon: Ear of Wheat (John 12:24) (the promised Messiah)
>Almurredin: Comes to Reign (spiritual reign)
>Porrima: The Coming Prophet (or Proclaimer or Prophecy Fulfilled)

Coma Berenices: The Desired of Nations
>Haggai 2:6–7
>For thus says the LORD of hosts: Once again, in a little while, I will shake the heavens and the earth and the sea and the dry land; and I will shake all the nations, so that the desire of all nations shall come, and I will fill this house with splendor, says the LORD of hosts. (NRSV)

Coma is the picture of a woman with a child in her arms. In ancient Egypt, this decan was called Shesnu, which means "The Desired Son." In Hebrew, Coma means "The Desired" or "Longed For." The Desired Son is Jesus, the Messiah. The Hebrew form of the word *Coma* is used in Haggai 2:7 (above). Again, the woman represents the Jews, Israel. This is the desired and awaited coming of the Messiah by the Jews.

Stars in Coma and their meaning:
>Diadem: Crown (the crown He would win by His victory, which is death and resurrection, over Satan at His first coming, and wear at His second coming as King)
>Al Zimach: The Branch, The Seed (the first coming)

Centaurus: The Despised (the man of double nature in humiliation)
>Isaiah 53:3
>He was despised and rejected—a man of sorrows, acquainted with bitterest grief. We turned our backs on him and looked the other way when he went by. He was despised, and we did not care. (NLT).

Centaurus is a figure of a Being with two natures (representing the Christ: God and man), with a spear pointed at Lupus (representing Christ the victim in the next decan) to indicate that He laid down his own life. The Hebrew name for Centaurus is Bezeh, which means "The Despised." This word is used in Isaiah 53:3 to describe Jesus. Other names for the constellation include Asmeath, meaning "A Sin Offering," and Cherion, which is Greek and means "The Pierced" or "One Who Pierces." The dual-natured Being is over the cross in the decan of Libra, showing his own death and the nature of it. He was despised by the Jews because He didn't fulfill their dreams of redemption from physical oppression. He was despised by the gentiles because He came first for the Jews. And few understood the real meaning of His first coming. And he will be despised at His second coming because He does not come for the redemption of all.

Star in Centaurus and its meaning:
>Toliman (Alpha Centauri): The Heretofore and the Hereafter (the first and second coming)

Bootes: The Coming; the exalted shepherd and harvester (the gathering together of the two churches)
>John 10:16
>And other sheep I have, which are not of this fold: them also I must bring, and they shall hear my voice; and there shall be one fold, and one shepherd. (ASV).

Bootes is pictured as a man walking quickly with a spear in his right hand and a sickle in his left hand. Bootes comes from the Hebrew Bo, which means "The Coming." He came first as a shepherd (with a spear) to gather the flocks together and protect them from harm. He will come again to "harvest" the "fruit of His labor."

Stars in Bootes and their meaning:
>Arcturus: He Comes (the first and second coming)
>Al Katurops: The Branch, Treading under Foot (the first and second coming)
>Mirac or Mizar or Izar: The Coming Forth as an Arrow; the Preserver or Guarding (the first coming as a shepherd)
>Muphride: Who Separates (the second coming; to separate the sheep from the goats)
>Al Nekkar: The Pierced (the first coming; pierced for our transgressions)
>Al Merga: Who Bruises (the first coming, bruising the enemy, Satan)

Bootes is the shepherd holding a spear to rule and guard his flock and the harvester holding a sickle to gather His crop. His purpose is not singular. Bootes is not the hunter holding two dogs on leashes.

This grouping represents the first coming of the promised Redeemer.

THE SIGN LIBRA AND ITS DECANS

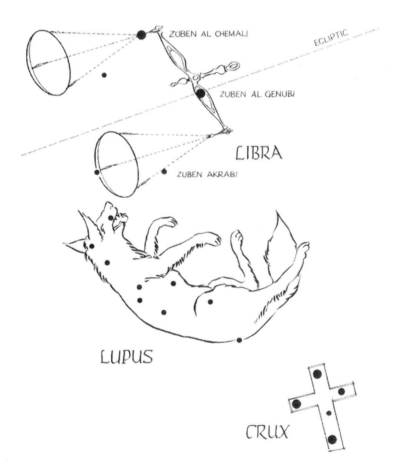

Reproduction from Albumasar's Zodiac. Public domain.

Libra: The Required Price Paid
> 1 Timothy 2:5–6
> For there is one God, and one mediator between God and men, the man Christ Jesus; Who gave himself a ransom for all, to be testified in due time. (NKJV).

Libra tells how Jesus paid the price to redeem us through His death on the cross and thereby gained the crown of glory. Libra is portrayed by a pair of scales. Scales were often used in trading to weigh the price required (and paid). In Arabic, the constellation was called Al Zubena, meaning "purchase" or "redemption." We see by the meaning of the names of the stars that man has tried to pay the required price on his own but could not. The only one who could pay the required price was Jesus.

Stars in Libra and their meaning:
> Zuben Al Genubi: The Price Which Is Deficient (man's inept attempt to pay for his own redemption)
> Zuben Akrabi: Redemption of the Conflict (the price paid through the conflict with the enemy)
> Zuben Al Chemali: The Price Which Covers (Jesus's blood sacrifice)
> Al Gubi: Heaped Up (His price was the highest price)

Crux: Cutting Off
> Daniel 9:26
> And after the sixty-two weeks Messiah shall be cut off, but not for Himself; and the people of the prince who is to come shall destroy the city and the sanctuary. The end of it shall be with a flood, and till the end of the war desolations are determined. (NKJV).

Crux, the Southern Cross, tells of the cross Jesus endured. In Hebrew, its name was Adom, which means "Cutting Off," which was referred to in Daniel 9:26 (above). This constellation was just visible above the horizon in Jerusalem at Jesus's first coming (and at His crucifixion) from the time of creation. It has not been seen in northern latitudes since then. This is due to the precession of the earth's axis as seen here:

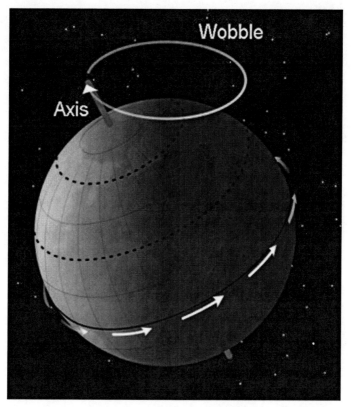

Precession. NASA image. Public domain.

In short, the earth's axis is not fixed but scribes a circle in space as in the above picture.

What this means is that our north pole will point at different stars through time as shown below.

Credit: Dennis Nilsson, Creative Commons Attribution 3.0 Unported.

So our "North Star" changes through time. Because of the tilt of the earth's axis, constellations will change "altitude" in the night sky through time. To make one complete cycle will take approximately 25,800 years. This will have significant meaning as we speak of time in relation to earth's history. I will speak more on this later.

Now, back to our story.

Lupus: The Victim Slain
> Revelation 13:8
> All who dwell on the earth will worship him, whose names have not been written in the Book of Life of the Lamb slain from the foundation of the world. (NKJV).

This picture shows an animal that is falling down dead. The modern name for this constellation is Lupus, which means "Wolf." It is also known as Victima, which is Latin for "The Victim" and Asedah, which is Hebrew and means "To Be Slain." Isaiah 53:7 says, "He was lead like a lamb to the slaughter . . ."

Remember Centaur's spear is pointed at Lupus, showing that Jesus laid down His own life. Note that Revelation 13:8 says, "from the foundation of the world." This speaks of creation. From day four of creation, the stars foretold of the coming sacrifice.

Corona: The Crown Purchased
> Hebrews 2:9
> What we do see is Jesus, who "for a little while was made lower than the angels" and now is "crowned with glory and honor" because he suffered death for us. Yes, by God's grace, Jesus tasted death for everyone in all the world. (NLT).

Corona is a picture of the crown of glory Jesus gained by His sacrifice as stated in Hebrews 2:9. In Hebrew, this constellation is called Atarah, which means "Royal Crown." Jesus made the supreme sacrifice—that is, He became a man and died a horrible death—just to gain a crown. The crown only represents what He gained. We can't begin to imagine His gain even as we can't begin to imagine His sacrifice.

Stars in Corona and their meaning:
> Al Phecca: The Shining
> Zuben Akrabi: The Gain Purchased

This grouping represents the price paid by Christ for the crown of glory, because man could not pay the required price even though he tried. (And men are still trying to pay the price on their own.)

THE SIGN SCORPIO AND ITS DECANS

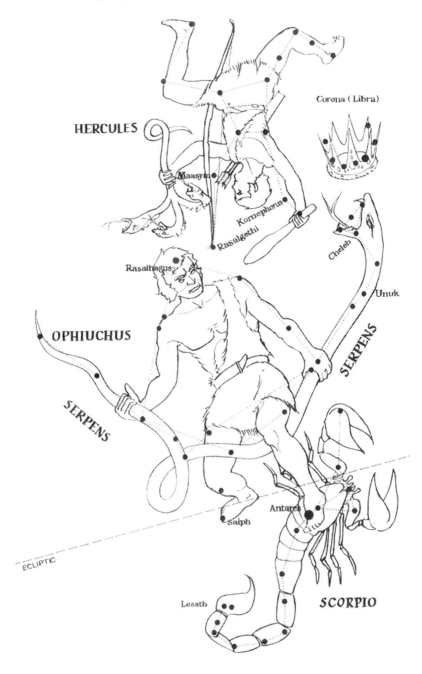

Reproduction from Albumasar's Zodiac. Public domain.

Scorpio: Wounding
> Isaiah 53:5
> But He was wounded for our transgressions, He was bruised for our iniquities; The chastisement for our peace was upon Him, And by His stripes we are healed. (NKJV).

Scorpio the scorpion is trying to wound a man's heel but is trodden under foot by that same man. Scorpio shows the fulfillment of Genesis 3:15, where God was speaking to the serpent after Adam's fall: "And I will put enmity between you and the woman, and between your offspring and hers; he will crush your head, and you will strike his heel."

Stars in Scorpio and their meaning:
> Antares: The Wounding (the "wounding" Christ would receive from the enemy)
> Leseth: The Perverse (in Scorpio's tail)
> Graffias: Holds or Holding (in Scorpio's claw as the "hold" Satan has on all of us)

Note: Is it any surprise that Antares, at the heart of Scorpio, is a red giant?

Serpens: The Conflict for the Crown
> Revelation 12:7–9
> Then there was war in heaven. Michael and the angels under his command fought the dragon and his angels. And the dragon lost the battle and was forced out of heaven. This great dragon—the ancient serpent called the Devil, or Satan, the one deceiving the whole world—was thrown down to the earth with all his angels. (NLT).

The serpent is struggling for dominion. This is the serpent trying to get the crown (Corona, a decan of Libra). This decan should always be described with the decan Ophiuchus because the two are locked in deadly conflict. The serpent is struggling for dominion, but Ophiuchus is restraining him. In this conflict, the two cannot be separated.

This is the serpent in Conflict for the Crown.

Stars in Serpens and their meaning:
> Unuk: Encompassing (the serpent is trying to wrap himself around Ophiuchus in the process to get the crown)
> Cheleb: The Serpent Enfolding (in the serpent's jaw)

Ophiuchus: The Evil One Held
> 2 Thessalonians 2:5–7
> Do you not remember that I told you these things when I was still with you? And you know what is now restraining him, so that he may be revealed when his time comes. For the mystery of lawlessness is already at work, but only until the one who now restrains it is removed. (NRSV).

Together with Serpens, this decan shows Ophiuchus (representing Christ) grasping and restraining the serpent (representing Satan) while stepping on the Scorpion's head and heart (Antares). The serpent is trying to get the Crown (a decan of Libra) to obtain the crown of dominion. The Greek name Ophiuchus is derived from the Hebrew Afeichus, which means "The Serpent Held."

Note that one foot of Ophiuchus is "crushing" the head of the scorpion while the scorpion's tail is pointed at his other foot to "bruise" his heel.

Stars in Opheuchus and their meaning:
> Ras al Hagus: The Head of Him Who Holds (in Ophiuchus's head)
> Triophas: Treading under Foot (Ophiuchus is treading on Scorpio)
> Saiph: Bruised (in the heel of Ophiuchus above Scorpio's sting)
> Carnebus: The Wounding (of Ophiuchus by the Scorpion)
> Megeros: Contending (the battle with the serpent)
> Sabik: Preceding

This is Christ restraining the serpent and struggling with the Enemy.

Hercules: The Mighty Victor
>2 Thessalonians 2:8
>And then the lawless one will be revealed, whom the Lord Jesus will destroy with the breath of his mouth, annihilating him by the manifestation of his coming. (NRSV).

This is Christ, humbled (on one knee) yet victorious over the enemy (one foot over the head Draco, the dragon, representing Satan, a decan of Sagittarius). He has a club in his right hand and a three-headed monster in his left. The skin of a lion that he has slain is thrown around him. Psalm 91:13 says of Christ, "The young lion and the dragon shall you trample under foot." (KJV).

Stars in Hercules and their meaning:
>Ras El Gethi (in his head): The Head of Him Who Bruises
>Kornephorus: The Branch, Kneeling (the humbled Messiah)
>Marsic: The Wounding (the Messiah wounding the enemy)
>Ma'asyn: The Sin Offering (the Messiah humbled by physical death)
>Guiam: Punishing (the Messiah punishing the enemy)

This grouping represents the battle Christ would fight.

Note: This group is unusual in that Scorpio is the zodiac sign on the ecliptic, although Ophiuchus actually shares the ecliptic with Scorpio. And pictorially, the sun will be in Ophiuchus longer than in Scorpio.

But, as we shall see, there is significance to this.

THE SIGN SAGITTARIUS AND ITS DECANS

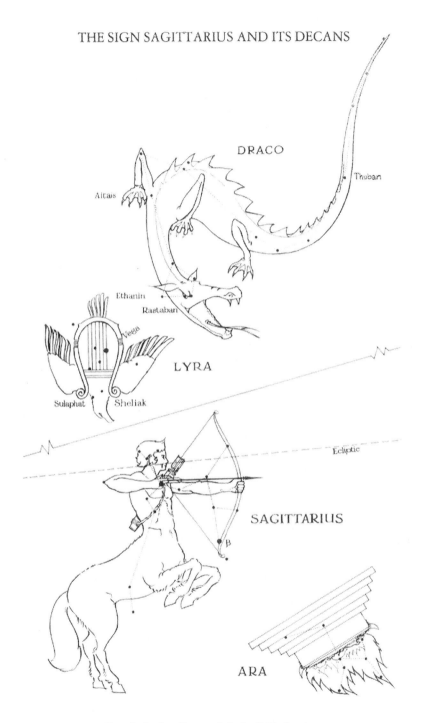

Reproduction from Albumasar's Zodiac. Public domain.

Sagittarius: The Archer, The Piercer, Prince of the Earth, Going Forth
> Psalm 45:5
> Your arrows are sharp, piercing your enemies' hearts. The nations fall before you, lying down beneath your feet. (NLT).

Sagittarius shows the dual-natured (God-man) Redeemer about to shoot an arrow through the enemy Scorpio's heart (in the previous sign). Sagittarius is the swift coming forth of the Redeemer (Prince of the Earth) as an arrow.

Stars in Sagittarius and their meaning:
> Naim: The Glorious One (the Messiah represented the glory of the Father)
> Al Shalua: The Dart (the swift sending forth of the Messiah)
> Al Warida: Who Comes Forth (the sending forth of the Messiah)
> Kaus: The Arrow (the swift sending forth of the Messiah)
> Ramih: Sent Forth, as an arrow (the swift sending forth of the Messiah)

Lyra: Praise for the Victorious Redeemer
> Psalm 148:13–14
> Let them praise the name of the LORD, for his name alone is exalted; his glory is above earth and heaven. He has raised up a horn for his people, praise for all his faithful, for the people of Israel who are close to him. Praise the LORD! (NRSV).

Lyra is a picture of a lyre or harp representing praise to the victorious Redeemer.

Star in Lyra and its meaning:
> Vega: He Shall be Exalted

Ara: Completing or Finishing
> Revelation 20:10
> Then the Devil, who betrayed them, was thrown into the lake of fire that burns with sulfur, joining the beast and the false prophet. There they will be tormented day and night forever and ever. (NLT).

Ara (the Altar) is the consuming fire prepared for His enemies. Ara is an upside-down altar with fire blazing downward, depicting the lake of fire prepared as the final destruction for the Devil and his followers.

Draco: Trodden On
> Psalm 91:13
> You will trample down lions and poisonous snakes; you will crush fierce lions and dragons under your feet! (KJV paraphrased).

Draco is the serpent, a great dragon (Satan) cast out and trodden on (by Hercules).

Stars in Draco and their meaning:
> Rastaban: The Head of the Subtle
> Ethanin: The Long Serpent
> Thuban: The Subtle
> Giansar: The Punished Enemy
> El Atik: The Fraudulent

Note: Draco's head is under Hercules's foot.

This grouping represents the victory Christ would win over His enemy, Satan, in His first coming.

The following verse shows this sign and its decans all in the same section:

Psalm 21:9, 11–13
> You will destroy them as in a flaming furnace when you appear. The LORD will consume them in his anger; His fire will devour them . . . Although they plot against you, their evil schemes will never succeed. For they will turn and run when they see your arrows aimed at them. We praise you, LORD, for all your glorious power. With music and singing we celebrate your mighty acts. (NLT).

Verse 9 shows Ara's fire against His enemies; verse 12 shows Sagittarius's arrow; verse 13 shows Lyra's praise for the Redeemer.

The second grouping shows the result of the Redeemer's work, the blessings to the redeemed. The constellations include Capricorn, The Atoning Sacrifice and Resurrection; Aquarius, The Holy Spirit Poured upon the Redeemed; Pieces, The Redeemed Blessed though Bound; and Aries, The Crowned Lamb and His Bride.

THE SIGN CAPRICORN AND ITS DECANS

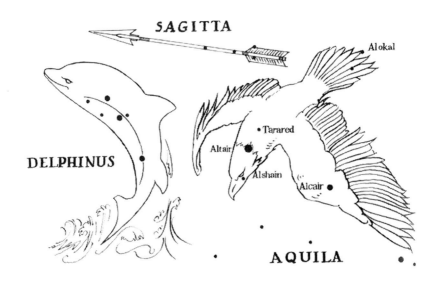

Reproduction from Albumasar's Zodiac. Public domain.

Capricorn: Life out of Death
>Romans 5:10–11
>For if, when we were enemies, we were reconciled to God by the death of his Son, much more, being reconciled, we shall be saved by his life. And not only so, but we also joy in God through our Lord Jesus Christ, by whom we have now received the atonement. (NKJV).

Capricorn is a symbol of the sacrifice (death) and resurrection of Jesus and the blessings (life) procured for the redeemed. Capricorn is Latin for "goat" and is depicted by a dying goat having the tail of a live fish. It symbolizes the goat of atonement slain for the redeemed and new life (a new beginning for an old church, the Jews, and a new church, the Gentile flock) coming forth in multitudes (as a fish spawning). In Old Testament times, the goat was used as a sin offering to take away the guilt of the community by making atonement for them before the LORD.

Stars in Capricorn and their meaning:
>Deneb Al Gedi: Cut Off (the sacrifice)
>Dabih: The Sacrifice Slain (the sacrifice)
>Ma'asad: The Slaying, Destroying (the destruction of sin's grasp by the sacrifice)
>Sa'ad al Naschira: Who Carries (He who carried away the grasp of sin)

Sagitta: Arrow
>Psalm 38:2
>For your arrows have pierced me, and your hand has come down upon me. (NKJV).

Sagitta, the Arrow, is the Arrow of God sent forth. This arrow is not meant for the enemy but for the Son of God, and it is sent by God. This is also seen in the prophecy about Jesus Christ in Psalm 38. Man did not crucify Jesus because it was ordained by God from before time.

Star in Sagitta and its meaning:
>Al Sahm: Arrow (God's arrow sent forth for His Son)

Aquila: The Pierced One Falling
> John 19:34, 36–37
> One of the soldiers, however, pierced his side with a spear, and blood and water flowed out . . . These things happened in fulfillment of the Scriptures that say, "Not one of his bones will be broken," and "They will look on him whom they pierced." (NLT).

Aquila represents Jesus Christ, the pierced one, falling in death.

Stars in Aquila and their meaning:
> Altair: The Wounding (of Christ)
> Al Cair: The Piercing (of Christ)
> Tarared: Wounded or Torn (of Christ)
> Al Shain: Bright (describes Christ) (Red Giant)
> Al Okal: Wounded in the Heel (Genesis 3:15 fulfilled)

Delphinus: Resurrection from Death
> Matthew 28:6
> He is not here; for He is risen, as He said. Come, see the place where the Lord lay. (NKJV).

Delphinus is a dolphin representing life and is connected to, actually emerging from, the goat's tail, representing sacrifice and death. Many ancient planispheres show Delphinus connected to Capricorn's tail by a stream of water.

Stars in Delphinus and their meaning:
> Sualocin: Swift (the swift emergence of life from death)
> Rotanev: Swiftly Running (the swift process of life from death)

This grouping represents the blessing of life (of the believer and a new church) from death (of the Redeemer).

THE SIGN AQUARIUS AND ITS DECANS

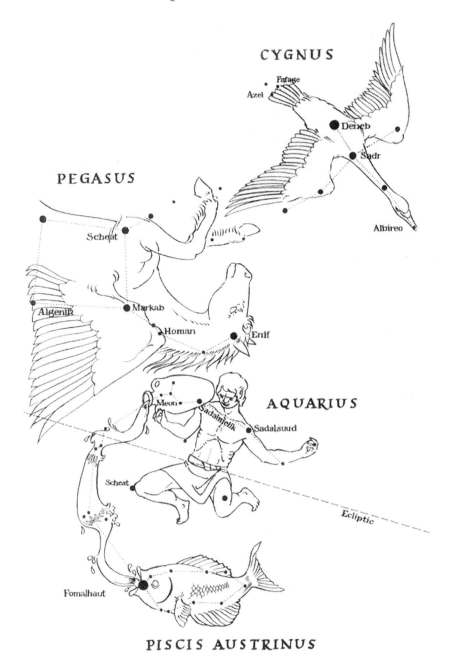

Reproduction from Albumasar's Zodiac. Public domain.

Aquarius: Blessing out of Victory
> John 4:10
> Jesus answered and said to her, "If you knew the gift of God, and who it is who says to you, 'Give Me a drink,' you would have asked Him, and He would have given you living water." (NKJV).
>
> Isaiah 44:3
> For I will pour water on the thirsty land, and streams on the dry ground; I will pour out my Spirit on your offspring, and my blessing on your descendants. (NRSV).
>
> Joel 2:28–29
> And it shall come to pass afterward, that I will pour out my spirit upon all flesh; and your sons and your daughters shall prophesy, your old men shall dream dreams, your young men shall see visions: And also upon the servants and upon the handmaids in those days will I pour out my spirit. (ASV).

Aquarius is a symbol of the giving of the Holy Spirit to God's people. Aquarius in Latin means "pourer forth of water." Aquarius is a man pouring water from an urn into the mouth of a fish. The water represents the Holy Spirit and the fish represents God's people, the general church. The Holy Spirit was given at Pentecost, fifty days after the resurrection of Jesus. It also represents the "pouring out of the Spirit" in the "last days," which is seen in the name of one star.

Stars in Aquarius and their meaning:
> Sadal Suud: He Who Pours Out (the first and second coming)
> Sheat (Skat): Who Goes and Returns (the second coming)
> Meon: The Urn (the Father, the source of the blessing)

Piscis Austrinus: Living Water Received
> John 4:15
> The woman said to Him, "Sir, give me this water, that I may not thirst, nor come here to draw." (NKJV).

THE FUTURE IS IN THE STARS

Acts 1:4–5
In one of these meetings as he was eating a meal with them, he told them, "Do not leave Jerusalem until the Father sends you what he promised. Remember, I have told you about this before. John baptized with water, but in just a few days you will be baptized with the Holy Spirit." (NLT).

Star in Piscis Austinus and its meaning:
Fomalhaut: Mouth of the Fish (the believers receiving the blessing)

Pegasus: The Soon Return of the Redeemer
James 5:7–8
Dear brothers and sisters, you must be patient as you wait for the Lord's return. Consider the farmers who eagerly look for the rains in the fall and in the spring. They patiently wait for the precious harvest to ripen. You, too, must be patient. And take courage, for the coming of the Lord is near. (NLT).

Revelation 22:20
"Surely I come quickly." (KJV).

Pegasus is a winged horse. Pegasus is the combination of two words: Pega, meaning "Chief," and Sus, meaning "Swiftly Coming." Pegasus is a picture of the Redeemer coming back swiftly (and soon) for His people.

Stars in Pegasus and their meaning:
Markab: Returning from Afar (the second coming)
Scheat: Who Carries (His people away)
Enif: The Branch (the Messiah)
Matar: Who Causes to Overflow with Joy (the result of His coming)

Cygnus: The Sure Return of the Redeemer
John 14:3
And if I go and prepare a place for you, I will come again and receive you to Myself; that where I am, there you may be also. (KJV).

Cygnus is a swan flying across the heavens. In both Greek and Latin, its name means "circling" (as in going and returning) and "returning." In the zodiac of Dendereh, it was called Tesark, which means "This from Afar."

Stars in Cygnus and their meaning:
    Deneb: The Lord Comes (the second coming)
    Albireo: Flying Quickly (the swift and soon return)
    Sadr: Who Returns as in a Circle (the second coming)
    Azel: Who Returns Quickly (the second coming soon after the ascension)
    Fafage: Shining Forth ("every eye shall see Him")

This grouping represents the church age, the blessing of the Holy Spirit, and the return of the Redeemer. This is the only grouping that speaks of the end of the church age, the "last days" and the second coming exclusively. It speaks only of His actual return with no mention of Satan or why He is returning (as absolute ruler).

THE SIGN PISCES AND ITS DECANS

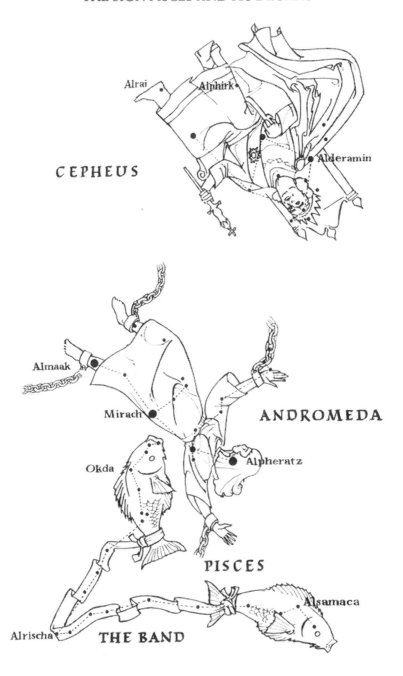

Reproduction from Albumasar's Zodiac. Public domain.

Pisces: Deliverance out of Bondage
>Colossians 1:13
>He has delivered us from the power of darkness and conveyed us into the kingdom of the Son of His love. (NKJV).

Pisces is two fish bound together by a band (a separate constellation) attached to Cetus (a sea monster representing Satan, a decan of Aries, representing Christ, which is holding or supporting the Band). This is a picture of the church restrained from freedom and blessing by being tied to Cetus but being supported by Christ. The two fish represent the "former" and the "latter" church or the "first" and the "last" church (the Jewish church and the church of Jew and Gentile). More on this will be seen in the constellation Ursa Minor and Ursa Major.

The Egyptian name from the zodiac of Dendereh is Picot Orion or Pisces Hori, meaning "The Fishes of Him That Comes."

Stars in Pisces and their meaning:
>Alsamaca (in the fish pointing outward): The Upheld (the whole church upheld by Christ)
>Okda (in the fish pointing upward): The United (the uniting of the former church to the latter church)

Note: Modern astronomy has the Band as part of Pisces. It is actually separate.

The Band: The Band
>John 10:16
>I have other sheep that do not belong to this fold. I must bring them also, and they will listen to my voice. So there will be one flock, one shepherd. (NRSV).

The Band that connects the two fish is not only being held by Cetus, but the Band is also binding and restraining Cetus, which is being supported by Aries. Explaining the Band is a little tricky. The Band is not so much a physical thing like the other constellations but more an abstract idea. What is it that joins the two "churches" together? It is the "Spirit of Christ." It is the "unity" of Christ held between believers, no matter who they are.

Star in the Band and its meaning:
> Al Risha: The Band or Bridle (that unites the former church (Jews) with the latter church (the general church)

Andromeda: The Redeemed in Bondage and Affliction
> Romans 8:21–22
> All creation anticipates the day when it will join God's children in glorious freedom from bondage. For we know that all creation has been groaning as in the pains of childbirth right up to the present time. (NLT).

Andromeda is the redeemed in their bondage and affliction. Andromeda's arms and feet are chained in misery and trouble. The chained woman represents Christ's church (His bride) in bondage to sin. In Hebrew, it is called Sirra, which means "The Chained." She awaits her release from this bondage.

Stars in Andromeda and their meaning:
> Alpheratz: The Broken Down (the church is broken down from bondage)
> Adhil: The Afflicted (the church is afflicted in bondage)
> Mirach: The Weak (the church is weak from bondage)
> Al Mara: The Afflicted

The Deliverer (Cepheus) will free her, as in Isaiah 52:1–3: "Loose yourself from the bands of thy neck, O captive daughter of Zion. For thus says the Lord, You have sold yourselves for nothing; and you shall be redeemed without money."

Cepheus: The Branch, the Deliverer Coming to Rule
> Revelation 19:15
> From his mouth comes a sharp sword with which to strike down the nations, and he will rule them with a rod of iron; he will tread the wine press of the fury of the wrath of God the Almighty. (NRSV).

Cepheus is the Crowned King, the Redeemer Coming to Rule. Cepheus is a glorious king, crowned, enthroned with a scepter in his hand. Cepheus means "the Branch." In the zodiac of Dendereh, it was called Pekahor, which means "This One Cometh to Rule."

Stars in Cepheus and their meaning:
> Alphirk: The Redeemer (coming to free the church)
> Al Deramin: Coming Quickly (the second coming will be swift and soon)
> Al Rai: Who Bruises or Breaks (the Redeemer breaking the bonds of the bound church)

This grouping represents the afflicted church waiting for the soon coming Redeemer.

THE SIGN ARIES AND ITS DECANS

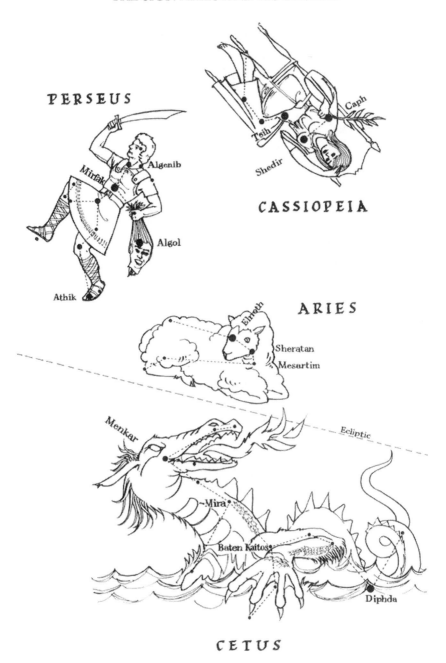

Reproduction from Albumasar's Zodiac. Public domain.

Aries: Glory out of Humiliation
>Revelation 5:12
>Saying with a loud voice: "Worthy is the Lamb who was slain to receive power and riches and wisdom, and strength and honor and glory and blessing!" (NKJV).

Aries is a ram or lamb, a symbol of the sacrificial ram or lamb. In Latin, Aries means "the Lamb" as the sacrifice, "the Chief" or "the Head" as the "head" of the church. Remember Abraham and the ram God provided as the sacrifice instead of Isaac.

Stars in Aries and their meaning:
>Al Nath: The Wounded (for our transgressions)
>Al Sheratan: The Bruised (for our iniquities)
>Mesartim: The Binding, the Bound (so there could be no escape)
>Al Botein: The Treading under Foot (even in death, the sacrifice was victorious)
>Al Thuraiya: The Multitude, the Abundance (the sacrifice was for the abundant multitude)

When Israel was told to prepare the Passover (the last plague on Egypt) before leaving Egypt, the sun had shortly before entered into the constellation Aries (more on this later).

When Jesus was crucified on the sixteenth day of the Jewish month Nisan (April 3), the sun was in Aries at the place in the stars between El Nath, meaning "Wounded," and Al Sheratan, meaning "The Bruised" (more on this later also).

Cassiopeia: Freed and Preparing for Her Marriage to the Redeemer
>Revelation 19:7
>For the time has come for the wedding feast of the Lamb, and his bride has prepared herself. (NLT).

Cassiopeia is the enthroned woman, the Bride of Christ, the church (God's people). Cassiopeia is the same woman who was chained in Pisces (Andromeda) but is now freed and enthroned, preparing for the marriage to her Husband, the Redeemer. She is fixing her hair with her left hand and adjusting her robe with her right hand to be "perfect" for her Husband. In Arabic, Cassiopeia's name is Ruchba, which means "The Enthroned" or "The Beautiful." In the zodiac of Dendereh, it is known as Set, meaning "Set Up As Queen." Another ancient name meant "The Daughter of Splendor."

Note that Cassiopeia is in the process of preparing; she is not yet ready. She has been freed from her chains of bondage as a result of the Redeemer coming and is just in preparation for the marriage. In old Jewish culture, when the "bride to be" was chosen and pledged to be married, a contract was drawn up that set the "bride price" that the groom would have to pay for his bride. Sometimes, if the groom had enough to pay the "bride price" at the time of betrothal, or part of it, he would. She would remain in her father's house (in bondage) while the groom went to prepare a place for them. When all was ready, he would return for her, pay any remaining "bride price" to her family, she would quickly get ready, and they would go to the wedding feast for marriage.

Christ (the groom) paid the full "bride price" (His sacrifice) at the time of betrothal before returning for the bride (the church). And all individual members get a down payment on the whole "bride price" when they become part of the church. This is what Ephesians 1:13 and 14 means.

> You, too, have heard the word of truth, the gospel of your salvation. When you believed in him you were sealed, (came under contract), with the promised Holy Spirit which is the guarantee or pledge of our inheritance until the redemption of the purchased possession (the church, the bride), unto the praise of his glory. (ISVNT).

Stars in Cassiopeia and their meaning:
   Shedar: The Freed (the redeemed church)
   Caph: The Branch (the One who frees)

Cetus: The Great Enemy Subdued and Bound
   Revelation 20:2
   He seized the dragon—that old serpent, the Devil, Satan—and bound him in chains for a thousand years. (NLT).

Cetus is a great sea monster, Satan, but he is the great enemy bound. Cetus is the largest of the constellations. Though the largest, not the most powerful. Aries is positioned above the head of Cetus, denoting authority over him, and has one foreleg extended onto the back of Cetus, representing the breaking of his power. This is the same leg supporting the Band.

Stars in Cetus and their meaning:
> Mira: The Rebel (he rebelled against God). Note: Is it any surprise that Mira, at the heart of Cetus, is a red giant?
> Menkar: The Bound or Chained Enemy (he will be chained (bound) for a thousand years at the second coming)
> Diphda: The Overthrown (he will be overthrown in the end)

Perseus: The Breaker Delivering the Redeemed
> Micah 2:12–13
> I will surely gather the remnant of Israel; I will put them together as the sheep of Bozrah . . . The breaker has gone out before them, they have broken through, they pass through the gate, they go out through it, and their king passes on before them, Jehovah at their head! (ASV paraphrased).

Perseus is "The Breaker" delivering the redeemed bride, the church. Perseus is a soldier with a sword in his right hand and a helmet on his head. In his left hand, he carries his adversary's head dripping with blood. Perseus is Latin for "The Breaker." In Hebrew, it is Peretz, also meaning "The Breaker." In the zodiac of Dendereh, Perseus was known as Kar Knem, meaning "Who Fights and Subdues."

Stars in Perseus and their meaning:
> Algenib: Who Carries Away (He will carry away the bride)
> Mirfak: Who Helps (He helps the bride)
> Atik: Who Breaks (He breaks the bounds, the chains of the bride)
> Algol: The Demon (representing Satan)

Note: Algol is located in the adversary's head, which Perseus is carrying. It is a variable star, which means that it varies in brightness (it gets dim, then bright, then dim, then bright, and so on). Algol is one of the best known eclipsing binaries, the first such star to be discovered, and one of the first non-nova variable stars to be discovered. Algol is actually a three-star system in which the large and bright primary is regularly eclipsed by the dimmer secondary companion. Thus, Algol's magnitude (brightness) is usually near constant but regularly dips by a magnitude of 1.3 every three days. There is also a secondary eclipse when the brighter star occults the fainter secondary. This secondary eclipse cannot be seen visually.

What is the significance of the makeup of this system? The larger, brighter, primary star is Jesus, whom Perseus, "The Breaker," represents. The smaller, dimmer, secondary star is Satan. Every three days (the time Jesus was in the grave), Satan *eclipses* Jesus for a short time and then yields again to the primary.

In astrology, Algol is considered the most unfortunate star in the sky.

This grouping represents the blessing realized.

The second coming of the Redeemer is shown by the third group of constellations. The constellations include Taurus, The Coming Judge of the Earth; Gemini, The Dual Natured King; Cancer, His Possessions Held Secure; and Leo, The Rending Lion of the Tribe of Judah.

THE SIGN TAURUS AND ITS DECANS

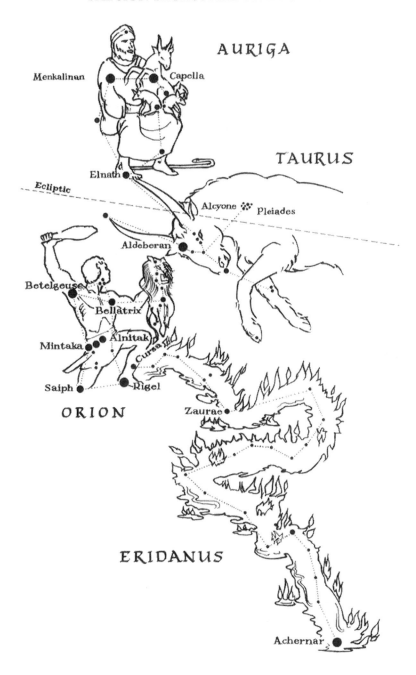

Reproduction from Albumasar's Zodiac. Public domain.

Taurus: The Judge
>1 Peter 4:5
>They will give an account to Him who is ready to judge the living and the dead. (NKJV).

>Jude 14–15
>Now Enoch, who lived seven generations after Adam, prophesied about these people. He said, "Look, the Lord is coming with thousands of his holy ones. He will bring the people of the world to judgment. He will convict the ungodly of all the evil things they have done in rebellion and of all the insults that godless sinners have spoken against him. (NLT).

Taurus, a ferocious bull called rimu, meaning "Wild Ox" in the Hebrew Scriptures, portrays Jesus's second coming to judge the earth. The wild bull symbolizes power and rule. In Numbers 23:22, Balaam described God's power in leading Israel: "God brought them out of Egypt; they have the strength of a wild ox (Hebrew: rimu)."

Stars in Taurus and their meaning:
>Aldeberan: The Governor (the absolute ruler)
>Elnath: The Wounded (a variation of Al Nath in Aries, but the same meaning, He who was wounded is now ruler)
>Alcyone: The Center or Central One (another ancient meaning is "Foundation Established")

Note: Alcyone is the "central" star of the star cluster known as "The Pleiades," meaning "Congregation of the Judge." Remember in Job 38:31 that God asks Job, "Can you bind the cluster of the Pleiades?" God is not only naming the cluster as we still know it, but He is stating an astronomical truth: the Pleiades stars are "bound" by gravitational forces to one another.

While the vast majority of stars in clusters move in different directions and thereby move away from one another, the stars in the Pleiades all move in the same direction at the same speed. And interestingly, it is estimated that the total number of stars in the cluster number in the thousands (as Enoch prophesied). There are many star systems, which means that the whole group is comprised of not only individual stars but also of groups within the group. Also, the cluster is located at the heart of Taurus. Is this a coincidence? I think not.

Also, Alcyone is an unusual star system. Alcyone is the primary with three stars orbiting it. The three stars are in their own orbits, whereas other systems have a primary with secondary stars orbiting it with other stars orbiting the secondary. What is the significance of the Alcyone system? Alcyone is God. The three orbiting stars represent the Trinity, the Three in One.

Orion: The Triumph and the Brightness of His Coming
> 2 Thessalonians 2:8
> And then the lawless one will be revealed, whom the Lord will consume with the breath of His mouth and destroy with the brightness of His coming. (NKJV).
>
> Isaiah 60:1–2
> Arise, shine; for your light is come, and the glory of the Lord is risen upon you. For, behold, the darkness shall cover the earth, and gross darkness the people: but the LORD shall arise upon you, and his glory shall be seen upon you. (KJV paraphrased).

Orion is The Glorious One coming. Orion is a mighty hunter (a picture of Jesus) with a big club in his right hand and a lion skin in his left hand. His left foot is raised to kill Lepus, his enemy (a decan of Gemini). Orion means "Coming Forth as a Light." In Hebrew, this constellation is called Chesil, meaning "A Strong One, a Hero." The Egyptians called it Hagat, meaning "This Is He Who Triumphs." The hilt of Orion's sword is the figure of a Lamb. Orion is a symbol of the Lord, the Light coming forth (the second coming).

Stars in Orion and their meaning:
> Betelgeuse: The Coming of the Branch (Jesus coming)
> Rigel (in the raised foot): The Foot that Crushes (to destroy the enemy)
> Bellatrix: Quickly Coming or Swiftly Destroying (the swift, and soon coming)
> Saiph: Bruised (bruised but returning in power)
> Alnitak: The Wounded One (wounded but returning in power)

Mintaka: Dividing as a Sacrifice (the sacrifice divides the purchased from the non-purchased)
Algebor: The Mighty (He cannot be defeated)
Alnagjed: The Prince (of Peace coming to rule)
Alnilam (Almirzam): The Ruler (absolute ruler)
Al Heka: Coming (He will return)

Eridanus: The Wrath of God Poured out on His Enemies (The River of the Judge)
Revelation 16:1
Then I heard a loud voice from the temple saying to the seven angels, "Go and pour out the bowls of the wrath of God on the earth." (NKJV).

Eridanus is the wrath of God poured out on His enemies. Eridanus is a river of fire that wanders through the sky toward Cetus (representing Satan and a decan of Aries), through the northern hemisphere, and ending at the star Achernar, meaning "The After Part of the River" in the southern hemisphere, after flowing under the heart of Cetus.

Stars in Eridanus and their meaning:
Achernar: The Afterpart of the River (when all is said and done)
Cursa: Bent Down (the beginning of the river over the head of Lepus, representing Satan)
Zourac: Flowing (near the central part of the river)

Auriga: The Protection of the Redeemed in the Day of Wrath
Ezekiel 34:22
God speaks of His flock: "I will save my flock, and they will no longer be plundered. I will judge between one sheep and another." (NRSV paraphrased).

Isaiah 40:11
He will feed his flock like a shepherd. He will carry the lambs in his arms, holding them close to his heart. He will gently lead the mother sheep with their young. (NLT).

THE FUTURE IS IN THE STARS

Auriga is the protection of the redeemed in the day of wrath. Auriga is a picture of a shepherd holding a mother goat with a pair of kids—safe and protected in the arms of the shepherd.

Stars in Auriga and their meaning:
    Capella: She Goat (the protected church, the mature)
    Men Kalinan: The Chain of the Goats (the protected church and its offspring altogether)
    Elnath: Wounded or Slain (as in John 10:11: "I am the good shepherd. The good shepherd lays down his life for the sheep.")
    Haedi: The Kids (the church, the young)

This grouping represents Christ's second coming as judge of the earth and the protected church from God's wrath at His coming.

THE SIGN GEMINI AND ITS DECANS

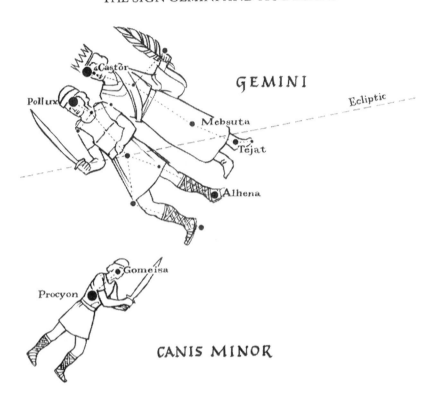

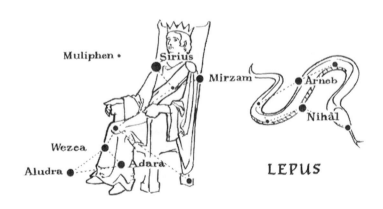

Reproduction from Albumasar's Zodiac. Public domain.

Gemini: His Rule on Earth
>Revelation 20:6
>Blessed and holy are those who share in the first resurrection. For them the second death holds no power, but they will be priests of God and of Christ and will reign with him a thousand years. (NLT).
>
>Philippians 2:6–7
>Though he was God, he did not demand and cling to his rights as God. He made himself nothing; he took the humble position of a slave and appeared in human form. (NLT).

Gemini symbolizes the dual nature of the Messiah: Suffering Servant and Conquering King. The twins of Gemini were called Castor and Pollux in Latin, though those are really the names of the stars in the head of each of the twins. The picture of twins represents the picture of two united as one. Gemini represents the two comings of Christ.

Stars in Gemini and their meaning:
>Castor: Ruler, Judge (the second coming)
>Mebsuta (in the leg of Castor): Treading Underfoot (the second coming)
>Pollux: Who Comes to Labor or Suffer (the first coming)
>Al Hena (in the foot of Pollux): Hurt, Wounded, or Afflicted (the first coming)
>Wasat: (in the waist of Pollux): Appointed
>Propus: The Branch, Spreading (both comings)

Lepus: The Enemy Trodden Underfoot
>Revelation 20:10
>The devil, who deceived them, was cast into the lake of fire and brimstone where the beast and the false prophet are. And they will be tormented day and night forever and ever. (NKJV).

Lepus is the enemy (Satan) trodden underfoot. Lepus is a serpent (pictured in later zodiacs as a hare, another astrology perversion of the story) that is located below Orion's foot (the foot that contains the star Rigel, "The Foot That Crushes").

Stars in Lepus and their meaning:
>Arneb: The Enemy of Him Who Comes (the enemy of Christ)
>Nibal: The Mad (going against all odds)
>Sugia: The Deceiver (the *great* deceiver)
>Rakis: The Bound (for a thousand years and then forever)

Canis Major: The Coming Prince of Glory
>Acts 5:31
>Then God put him in the place of honor at his right hand as Prince and Savior. He did this to give the people of Israel an opportunity to turn from their sins and turn to God so their sins would be forgiven. (NLT).

Canis Major is the Coming Prince of Glory. The modern picture of Canis Major is a dog (another astrological perversion). Older planispheres represent Canis Major as Naz, "The Hawk Coming Swiftly Down." The most ancient planispheres have Canis Major as a prince on a throne. Canis Major represents the second coming.

Stars in Canis Major and their meaning:
>Sirius: Prince (the ruling prince will bring about peace)
>Mirzam: Ruler (absolute ruler)
>Adara: The Glorious (the glorious ruler, ruling in wisdom)
>Wezea: The Bright, The Shining (ruling with justice)
>Alshira Aljemeniya: The Prince of the Right Hand (absolute rule)

Canis Minor: The Coming Redeemer
>Job 19:25
>For I know that my redeemer lives, and that he shall stand at the latter day upon the earth. (NKJV).

Canis Minor is "The Coming Redeemer." The modern picture of Canis Minor is a dog (yet another astrological perversion) smaller than Canis Major. The ancient Egyptians called it Sebak, "The Conquering" or "Victorious." Its ancient picture was a human with a hawk's head. Canis Minor represents the first coming.

Stars in Canis Minor and their meaning:
> Gomeisa: Bearing for Others (having taken the sin of the world)
> Procyon: Redeemer, Savior (He paid the required price)
> Gomeisa: Burdened or Bearing for Others (having taken the sin of the world)
> Alshira Alshemeliya: The Prince of the Left Hand (ruling in peace)

This grouping represents the first coming of Christ as a spiritual ruler, the Prince of Peace, and the second coming of Christ as absolute ruler, to whom every knee shall bow, ruling in authority, wisdom, and power. He paid the required price to gain the crown worthy of glory.

THE SIGN CANCER AND ITS DECANS

Ursa Minor

Ursa Major

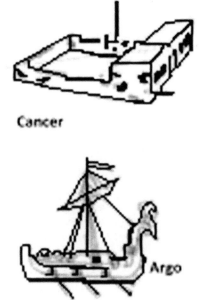

Cancer

Argo

Reproduction from Albumasar's Zodiac. Public domain.

Cancer: His Possession Held Secure
> 1 Peter 1:3–5
> Blessed be the God and Father of our Lord Jesus Christ! By his great mercy he has given us a new birth into a living hope through the resurrection of Jesus Christ from the dead, and into an inheritance that is imperishable, undefiled, and unfading, kept in heaven for you, who are being protected by the power of God through faith for a salvation ready to be revealed in the last time. (NRSV).

Cancer is a symbol of a shelter for God's people. Cancer comes from a Latin root word meaning "to hold or encircle." In the zodiac of Denderah, it was called Klaria, meaning "cattlefolds."

The modern picture for Cancer is a crab. From the translations of its name and the names of its stars, it very likely had a different ancient picture.

Stars in Cancer and their meaning:
> Tegmine: Holding (safely)
> Acubene: Sheltering or Hiding Place (a place of safety)
> Ma'alaph: Assemble Thousands (the church)
> Alhimarean: The Kids or Lambs (the church)

Note: There is a star cluster (with nebula) in Cancer called Praesepe (the Manger). This cluster is visible to the naked eye. There are two stars, one on either side of the cluster: Asellus Borealis and Asellus Australis, which are The Northern Donkey Colt and The Southern Donkey Colt, respectively.

Ursa Minor: The Stronghold of the Saved
> John 10:27–28
> My sheep recognize my voice; I know them, and they follow me. I give them eternal life, and they will never perish. No one will snatch them away from me . . . (NLT).

Ursa Minor is The Stronghold of the Saved. Ursa Minor is now known (improperly so) as the Little Bear. Another name for it was Arcas, meaning "The Stronghold of the Saved." It is likely that the Greeks mistakenly thought the constellation was called by the Persian word *dob*, meaning "bear," rather than the Hebrew word *dowb*, meaning "sheepfold." The word "little" should be "lesser." Thus, by the original meaning of this constellation, it should be a sheepfold (the lesser sheepfold), representing the original church of Jewish believers.

Stars in Ursa Minor and their meaning:
>Al Ruccaba (Polaris): The Turned (Revolved Around) (everything revolves around Christ)
>Kochab: Waiting for the Coming, Waiting Him Who Comes (the second coming)
>Pherkadain: The Calves or The Redeemed Assembly (the church)
>Yildun: Star (shining)
>Ahfa al Farkadain: Dimmer (Former) Calf (the former assembly, the Jews)
>Anwar al Farkadain: Brighter (Latter) Calf (the latter assembly, the whole church)

Ursa Major: The Assembled Flock
>John 10:27–28
>My sheep recognize my voice; I know them, and they follow me. I give them eternal life, and they will never perish. No one will snatch them away from me . . . (NLT).

>John 10:16
>I have other sheep that do not belong to this fold. I must bring them also, and they will listen to my voice. So there will be one flock, one shepherd. (NRSV).

Ursa Major is the Assembled Flock. Ursa Major is known as the Big Bear. Job 9:9 and 38:32 refer to this constellation. The Hebrew word translated as "bear" means "To Hasten or Assemble Together." The constellation is called Alnaish, meaning "The Assembled Together" by the Arabs. As with Ursa Minor, it is felt that the Greeks may have mistakenly thought the constellation was called by the Persian word *dob*, meaning "Bear," rather than the Hebrew word *dowb*, meaning "Sheepfold," hence, the pictures of the big bear and the little bear. The term "Big" comes from the word associated with sheepfold meaning "greater." Therefore, the proper designation of this constellation should be the "Greater Sheepfold." Ursa Major represents the church at large, the combined church of Jew and Gentile.

Stars in Ursa Major and their meaning:
> Dubhe: A Flock (the church)
> Merak: Flock (the church)
> Phad: Guarded or Numbered (the shepherd knows how many there are and he guards them)
> Megrez: Separated (the sheep in the fold and goats outside the fold are separated)
> Alioth: She Goat (the sacrifice)
> Mizar: Small (immature sheep)
> Alcor: The Lamb (mature sheep)
> Alcaid: The Assembled (the church gathered in purpose)
> Benetnash: Assembled daughters (the church gathered for praise)

Note: As I said, Ursa Major and Ursa Minor are now known as bears. However, no bear in existence today, or ever in the past, has had a long tail.

Argo: Travelers Safely Home
> 1 Peter 1:3–5
> Blessed be the God and Father of our Lord Jesus Christ! By his great mercy he has given us a new birth into a living hope through the resurrection of Jesus Christ from the dead, and into an inheritance that is imperishable, undefiled, and unfading, kept in heaven for you, who are being protected by the power of God through faith for a salvation ready to be revealed in the last time. (NRSV).

Argo is The Travelers Safely Home. Argo means "A Company of Travelers." Argo is the picture of a ship carrying the redeemed of the Lord to a safe harbor for protection.

Stars in Argo and their meaning:
> Canopus: The Possession of Him Who Comes (we belong to Him)
> Tureis: Possession (He owns us)

Soheil: The Desired (we desire to see Him)
Asmidiska: The Travelers Released (released from the bondage of sin)
Sephina: Abundance or Multitude (a multitude of believers)

This grouping represents the protection of the redeemed from the coming wrath of God.

THE SIGN LEO AND ITS DECANS

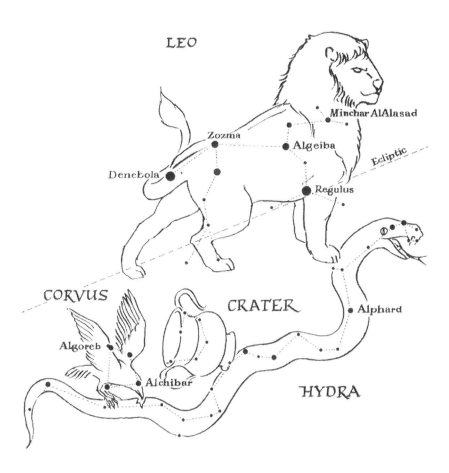

Reproduction from Albumasar's Zodiac. Public domain.

Leo: His Enemies Destroyed
> Hosea 5:14
> I will tear at Israel and Judah as a lion rips apart its prey. I will carry them off, and there will be no one left to rescue them. (NLT).

Leo is the Lion of the tribe of Judah. Leo is a lion standing on the serpent Hydra, representing his destruction.

Stars in Leo and their meaning:
> Regulus: The Rending Lion (He will utterly destroy the enemy)
> Denebola: The Judge Who Comes (He will come in judgment)
> Algiebha: The Exaltation (He will be exalted)
> Zozma: Shining Forth (His coming will be glorious)
> Minchar Alasad: The Tearing of the Lion (He will utterly destroy the enemy)
> Aldafera: Enemy Put Down (the enemy defeated)

In the zodiac of Dendereh, the word *Knem*, meaning "Who Conquers," is written under the constellation of Leo. In Hebrew, it was called Arieh, meaning "Lion." In Syriac, it was Aryo, meaning "The Rending Lion."

Hydra: The Abhorred
> Genesis 3:15
> "From now on, you and the woman will be enemies, and your offspring and her offspring will be enemies. He will crush your head, and you will strike his heel." (NLT).
>
> Revelation 20:10 speaks of this event:
> And the devil that deceived them was thrown into the lake of fire and brimstone . . . and shall be tormented day and night forever and ever. (NKJV).

Hydra is the picture of a serpent, "that old serpent, the Devil," being destroyed. And not just destroyed but tormented forever more. I will have more to say about Hydra with the next two decans.

Star in hydra and its meaning:
    Alphard: The Excluded, or Put out of the Way (Satan will be put "out of the way")

Crater: The Cup of Wrath
    Revelation 14:9–10
    Then a third angel followed them, saying with a loud voice, "If anyone worships the beast and his image, and receives his mark on his forehead or on his hand, he himself shall also drink of the wine of the wrath of God, which is poured out full strength into the cup of His indignation. He shall be tormented with fire and brimstone in the presence of the holy angels and in the presence of the Lamb. (NKJV).

Crater is a picture of a cup. Crater is the wrath of God poured out on the Serpent. It symbolizes the cup of the wrath of God on the His enemy. It is the cup of God's indignation.

Star in Crater and its meaning:
    Al Ches: The Cup (the cup of God's wrath)

Corvus: The Raven
    Revelation 19:17–18, 21
    Then I saw an angel standing in the sun, and with a loud voice he called to all the birds that fly in mid-heaven, "Come, gather for the great supper of God, to eat the flesh of kings, the flesh of captains, the flesh of the mighty, the flesh of horses and their riders—flesh of all, both free and slave, both small and great." . . . And the rest were killed by the sword of the rider on the horse, the sword that came from his mouth; and all the birds were gorged with their flesh. (NRSV).

Corvus is a raven who is tearing at the flesh of the serpent Hydra. Corvus is the devouring of the "Old Serpent."

Stars in Corvus and their meaning:
    Alchibar: Accursed (Satan is)
    Algoreb: Raven (the one who will devour the enemy)

> Minchar Algoreb: The Raven Tearing to Pieces (the destruction of Satan)
>
> Minchar Alsugia: The Deceiver Torn to Pieces (Satan destroyed)

This grouping represents the final judgment and eternal "destruction" of Satan and all those who have followed him.

Note: I want you to notice that in the groupings representing the second coming that there is no place in the signs of the zodiac to definitively indicate the "rapture." There are those which speak of the protection of the redeemed from the wrath of God. Do they speak of the rapture? I do not believe so. I do not mean to say that there will be no rapture; I believe the Bible speaks of it. My point is that the zodiac, and therefore God, did not place emphasis on it. The important thing is that man sinned and thereby gave Satan authority to rule on the earth (for a while). Jesus paid the required price to redeem (buy back) man, thereby breaking Satan's rule and authority. Jesus will return for His purchased church. Satan will in the end be forever put out of the way, and God will restore His creation to His original intent.

CHAPTER 7

The Galactic Equator

The meanings of the signs and their decans are all important, but there's more to make the future even more interesting. I must mention here the galactic equator. Just as the celestial equator is an imaginary line extending out through space from the earth's equator, dividing space symmetrically, so the galactic equator is an imaginary line cutting through our galaxy (the Milky Way), dividing the top and bottom (north and south), symmetrically (according to the official IAU galactic coordinate system). But this is not by a "horizontal" line, and I am in agreement with this.

The galactic plane is what divides the galaxy symmetrically in two horizontally (the galactic equator and galactic plane are not the same). There are twelve constellations which appear on the galactic equator. These constellations are not the signs of the zodiac, not the twelve signs on the ecliptic. (See diagram below: the purple line.) These twelve are of the decans of the zodiac. These twelve tell of the first and the second coming of Christ specifically. The first six are Cepheus, "The Branch"; Cygnus, "Coming, Going, and Coming Again"; Aquila, "The Pierced One"; Ophiuchus, "The Conqueror of the Serpent Receiving the Wound in the Heel"; Lepus, "The Victim, in the Hand of the Centaur (the conqueror)"; and the Southern Cross, "The Cutting off of the Last Adam." These are prophesies fulfilled.

The second six are Argo, "The Coming of the Desired with His People"; Canis Major, "The Prince"; Orion, "Coming Forth As Light" (or "The Brightness of His Coming"); Auriga, "The Shepherd"; Perseus, "The Breaker of Bonds"; and Cassiopeia, "The Enthroning." Could this be a coincidence or part of a divine design?

Note: This diagram shows the constellation boundaries as set by the International Astronomical Union. As stated earlier, I do not agree with their boundaries. The diagram also shows all of the eighty-eight constellations recognized by the IAU. Therefore, in this diagram, you will see the galactic equator pass through constellations I do not recognize for biblical astronomy purposes.

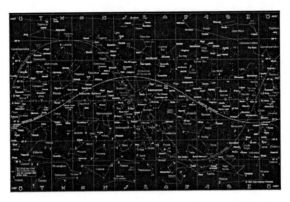

Copyright: Nick Anthony Fiorenza, lunarplanner.com. Image used with permission.

I want to look briefly at the phenomenon of our sun supposedly crossing the galactic equator on December 21, 2012, so widely publicized by "Doomsday Prophets" as a cosmic disaster. First, the ecliptic is on the same plane as the galactic equator, which means the sun is always on the galactic equator. The diagram below shows the motion of the sun supposedly across the galactic equator. It is actually the galactic plane.

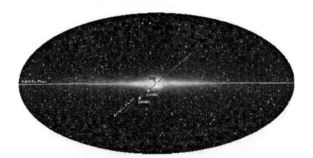

The Milky Way. NASA image. Public domain.

The following diagram, showing the galactic equator (as the "reference plane") and the galactic plane (as the "H1 plane"), is correct.

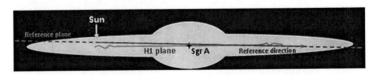

Credit: 2012hoax.org, Creative Commons Attribution 3.0 Unported.

The truth is the sun is "above" the galactic plane in space and about to cross that plane. What will happen when it does? No one knows. However, I will make a prediction: nothing will happen. My reasoning will be clear shortly. But for now, please don't be concerned with "the end of the world." After December 21, 2012, you can reread this and say, "He was right."

It may not seem like much at this point, but that's all we need to know of the galactic equator for now.

CHAPTER 8

Precession

I briefly mentioned precession earlier. I now want to look into the subject in more detail because this is where it starts to get interesting. First, let's define precession.

Precession. NASA image. Public domain.

Precession is the resulting "wobble" of the earth's axis in space (as seen in the above diagram) due to gravitational forces of the sun and other planets working on the earth. For those who aren't physicists, let's simplify this for our astronomical purposes. The position of the equinoxes is not fixed among the background of "fixed" stars but, due to precession, the equinoxes slowly shift westward over the ecliptic at a speed of just under one degree every seventy-two years. To give you an idea of how much this is, an average full moon spans a distance of about five degrees. So for the equinoxes to shift that distance would take 360 years. You and I will never notice the difference. This shifting effect is called precession of the equinoxes, because the equinoxes precess, or shift, their position on the ecliptic over time.

Note: Here again astrology is proven false. Precession of the equinoxes changes the sun's position in the constellations (astrological "signs" and even the constellations in the celestial coordinate system) at any one given day through time. Other precessional forces, which I won't get into here, are also at work. The precession of the earth's axis also causes changing pole stars. Polaris is our "north star" at present but has not always been, nor will not always be, so.

This precession causes the sun's length of stay in a constellation to change over time so the relation between the earth, stars, planets, and sun is not constant. Also, by which calendar should you choose your "birth sign"? Should you use God's original "Lunar-Solar calendar," where months were determined by the new moon, except for the addition of a day every three months? Or should you use the "Exodus" calendar which God gave to the Jews at the exodus where God changed the New Year? Or should you use the Gregorian calendar or the Julian calendar? The point is that astrology *cannot* account for these changes, but most people don't know about these things so, looking for hope somewhere, they believe astrology's lie.

Our astrological birth signs no longer coincide with the sun's location in the constellations of the zodiac because of precession. Even the Tropic of Cancer and the Tropic of Capricorn are now the Tropic of Taurus and the Tropic of Scorpio when we observe the sun on the solstices. Astrologers have neglected to update our present position in the cosmos for over 2,000 years. We are not the astrological birth signs the newspapers and a majority of the astrologers proclaim we are.

Now, back to the precession of the equinoxes. For the sun to travel full circle around the ecliptic, it would take about 25,800 years. Here's where it starts to get really interesting. Does the sun's position on the ecliptic mean anything? Astrology says it does. Is there any validity in that point? Yes, there is. But not in the way astrology says. Since there are no lines in the night sky to separate the constellations, how do we know where one ends and the next one begins? There are star charts showing the constellation boundaries, such as the following.

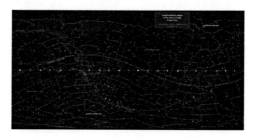

Copyright: Nick Anthony Fiorenza, lunarplanner.com. Image used with permission.

At present, according to the International Astronomical Union, there are eighty-eight constellations. The present constellation boundaries (shown here) were adopted by the International Astronomical Union in 1928 and published officially in 1930. However, there is a problem with this mapping. It is accepted in astronomical circles that Coma Berenices was invented by Gerardus Mercator in 1551 by taking a few stars from Leo and combining them with stars that were part of no existing constellation.

The problem is that Coma Berenices is one of the original forty-eight constellations which God set in the night sky, and even Ptolemy, in the second century AD, identified Coma Berenices as one of the original constellations. In other words, Gerardus Mercator was changing the gospel story told by the stars. Also, the IAU claims that stars from the claws of the constellation Scorpio were taken to become part of the constellation Libra. Also, in 1752, the French astronomer Nicolas Louis de Lacaille subdivided Argo into Carina (the keel, or the hull, of the ship), Puppis (the poop deck), and Vela (the sails). So this is another astrology inaccuracy: astrology uses the present constellations and their boundaries for its information and the present constellation formations do not represent the truth.

Every time the sun enters a new constellation due to the precession of the equinoxes, we are entering into a new astronomical age. According to the official IAU chart, we are presently in the "Age of Pisces" and will enter the "Age of Aquarius" around the year 2597. Given man's innate ability to mess up what God has given us, not to mention Satan's ability to "trick" man into believing a lie, I believe the constellation boundaries set by the International Astronomical Union are not accurate. If stars have been moved from one constellation into another, those boundaries cannot be right. If one constellation has been divided into four, those boundaries cannot be right. So by using a star that is central to the zodiac (the ecliptic north pole), named stars (the names God gave them) in the constellations, and the position of the sun on the ecliptic with respect to the precession of the equinoxes, I have defined constellation boundaries which I believe to be accurate from a biblical standpoint.

Before going further, let me say that I haven't said anything about the significance of the sun in the zodiac. In biblical astronomy, the sun represents the Father and His Glory. Its position in the signs is significant.

Since the constellations tell the gospel story, let's look at the constellation ages from a biblical perspective. What was happening when the sun (in relation to the equinoxes) was in any given constellation?

Here I need to talk just a little about the different calendars (periods of time that make up a year) that have been used throughout history. When God created the earth, sun, moon, and stars and set time in motion, the earth day was twenty-four of our hours (although Enoch reckoned the day to be eighteen equal units of time, and he had 365 years to verify it). The week was seven days. The month was twenty-nine days (the moon goes from new moon to new moon in twenty-nine days). The year then was what is now called a tropical year (the time it takes for the sun to travel along the ecliptic in relation to the four seasons). The tropical year is slightly shorter than the sidereal year we now observe (the time it takes for the sun to travel 360 degrees along the ecliptic in relation to the "fixed" stars of the zodiac).

The timeline I will use takes these differences into account, although those differences, amounting to only about sixty years, are not enough to really make any difference, unless you want to pinpoint the first day of creation. The important thing is the precession of the equinoxes match up with historical events and will give us the true age of the earth. Also, we must understand that from the time of creation to the exodus of Israel from Egypt, the "Creation Calendar" was in use. At the "exodus," God changed the New Year for the Jews to the month they left Egypt (Nisan on their calendar and March/April on ours). This is significant in relation to the precession of the equinoxes in relation to earth's historical timeline.

The following list gives the span of the constellations on the ecliptic in degrees and the corresponding length of time the sun is in each (not by the IAU boundaries).

Constellation	Degrees	Time in Years
Virgo	43.32	3,104.31
Leo	33.11	2,372.66
Cancer	12.29	880.70
Gemini	16.99	1,217.50
Taurus	24.22	1,735.60
Aries	20.03	1,435.34
Pisces	39.33	2,818.39
Aquarius	38.48	2,757.48
Capricorn	31.39	2,449.41
Sagittarius	33.29	2,385.56
Scorpio (Ophiuchus)	36.65	2,626.34
Libra	30.90	2,214.30
Total	360.00	25,997.59

CHAPTER 9

Precession and History

To begin our journey back in time, we will start form a point in time we know: the crucifixion of Jesus (there is enough reliable evidence available to be sure of its accuracy, plus one unusual source which I will discuss later). I have listed the Age of Aquarius first only because it is the latest date of significance.

Of course, I had to be sure this date corresponds to the zodiac clock/calendar. It does. The dates for the other events are derived from a comparison of multiple sources (very few of which are in agreement). I then checked those dates against the zodiac for verification. My date for creation matched one of the sixty-four proposed dates by "higher thinkers," noted Christian leaders, such as Martin Luther, and the early church fathers (of the first century). I also could not find any of my other proposed event dates listed by anyone else except for the exodus. That is not to say that they have not been listed by someone else at some time or another; that is only to say that in my limited search I could not find them. The location of the sun is derived from computer models for those dates.

Age of Aquarius	March 20, 2004 AD
Crucifixion of Jesus	April 3, 33 AD
Christmas	September 29, 2 BC
Exodus	March 28, 1453 BC
Flood	October 30, 2310 BC
Creation	October 14–20, 3996 BC

Before getting deeper into our explanation of the dates for the events listed, I need to say something about the planets and their role in biblical astronomy. First, they are just as much a part of the universe that God created as anything else, and God set them in place just as He did the stars. Their apparent position among the constellations has significant meaning just as their names have significant meaning. Their names can be traced back to the most ancient Hebrew language and beyond along with the names of the constellations and stars. And again, the usage of them in astrology is simply a perverted form of what God originally intended.

The planets not listed are not included in biblical astronomy because they are not visible to the naked eye. Uranus, at magnitude 5.8, is sometimes visible to us, depending upon atmospheric conditions, and would have been visible to the antediluvian patriarchs.

Since I have already mentioned that the sun represents God and His omnipotent glory, we should understand that the moon represents the Lord (Christ) and His ruling glory. The planets represent certain aspects of the Lord's glory. They are also associated with signs as their "houses."

Here are the planets, their meaning, and their "houses'" meaning in prophecy.

Saturn: Resting, as the work is complete and ready for a new beginning.
    Capricorn: the sacrifice slain (the completed work of the first coming; prepared for a new beginning).
    Aquarius: the water of purification (the blessing of the first work received; prepared for the second blessing).

Jupiter: The Lord Has Set Free; Righteous in Dominion.
    Sagittarius: deliverance by Him coming forth (first coming).
    Pisces: whose are the congregation (the whole church).

Mars: The Wounded (as Christ for His Church and His Church in Him).
    Aries: the lamb bruised (not killed).
    Scorpio: His heel bruised.

Venus: The Beloved; The Bright, Shining One; The Altogether Lovely; The Awaited One.
    Libra: redemption (the price paid in full by the sacrifice of the first coming).
    Taurus: deliverance of the righteous by faith; deliverance by judgment.

Mercury: Going and Returning Again; Power Coming.

THE FUTURE IS IN THE STARS

Virgo: the branch (the first and second coming in power, first as Redeemer, then as judge).
Gemini: the two comings.

Uranus: Ruling; Fearful Judgment.
Leo: the lion (the second coming as absolute ruling king).
Cancer: held secure (the righteous protected from the wrath of God's judgment).

The names and their meaning have come to us as the names of stars and constellations have: by ancient writings and tradition.

For each event, I will give the sign that the sun was in on that day. I will then give the constellations the planets were in and the constellation the moon was in that day. I will then list the sign the sun was in on the equinoxes and solstices that year. Then for the sign the sun was in on that day, I will give a Scripture and list stars in that sign appropriate to the event along with the decans for that sign, a Scripture, and their stars.

Here are the events:

Crucifixion of Jesus: Occurred on AD April 3, 33. The sun was in Aries.
Venus (The Beloved, The Awaited One) and Mercury (Going and Returning Again) were in conjunction in The Band (That Which Joins the Former and Latter Church) (which is Jesus). Venus, the aspect of Christ which the Jews would associate closer with was nearer to the "first" fish representing the "former" church, the Jews. Mercury, the aspect of Christ that (Gentile) "Christians" would associate with was nearer to the "second" fish representing the whole church of Jew and Gentile.
Jupiter (The Lord Has Set Free) and Mars (The Wounded) were in conjunction in Gemini (His rule on earth, the two comings) with the star Mebsuta (Treading under Foot). The wounded one (wounded by physical death), the Christ, was treading Satan underfoot by His death and resurrection while setting the "church" free from the bondage of sin.
Saturn (Resting, The Work Completed) was in Cancer. The "work" of redemption was complete. Now came the time for the church to be held secure (Cancer) until the time of the second coming (full redemption).

Uranus (Ruling, Fearful Judgment) was in Leo (the Ruling King). Although not the final rule of judgment, Satan was put in a place of fearful judgment because he had just lost this battle to await his final judgment.

The moon (the Lord's Glory) was in Virgo (The Seed) that morning, crossing into Libra (The Required Price Paid) that afternoon. The moon was near the star Spica (The Branch) in Virgo during the trial of Jesus, crossed the boundary into Libra between 9:00 a.m. and 3:00 p.m., when He was on the cross, and approached the star Zuben Al Genubi (The Price Which Is Deficient) (man's attempt to pay for his own redemption) by the time in the evening He was hurriedly buried.

On the spring equinox, the sun was in Pisces (near the "first" fish representing the Jews).

On the summer solstice, the sun was in Gemini (the two comings, His rule on earth).

On the fall equinox, the sun was in Virgo (the seed) near the star Spica (The Branch).

On the winter solstice, the sun was in Sagittarius (the Victorious One).

Aries: Glory out of Humiliation
>Revelation 5:12
>saying with a loud voice, Worthy is the Lamb who was slain to receive power and riches and wisdom, and strength and honor and glory and blessing! (NKJV).

Stars in Aries and their meaning:
>Al Botein: The Treading under Foot
>Al Thuraiya: The Multitude, The Abundance

Aries's decans:
>Cassiopeia: Freed and Preparing for Her Marriage to the Redeemer
>Cetus: The Great Enemy Subdued and Bound
>Perseus: The Breaker Delivering the Redeemed

Cassiopeia: Freed and Preparing for Her Marriage to the Redeemer
>Revelation 19:7
>For the wedding of the Lamb has come, and his bride has made herself ready. (NLT).

THE FUTURE IS IN THE STARS

Star in Cassiopeia:
    Caph: The Branch

Cetus: The Great Enemy Subdued and Bound
    Revelation 20:2
    He seized the dragon—that old serpent, the Devil, Satan—and bound him in chains for a thousand years. (NLT).

Star in Cetus:
    Diphda: The Overthrown

Perseus: The Breaker Delivering the Redeemed
    Micah 2:12–13
    "I will surely gather the remnant of Israel; I will put them together as the sheep of Bozrah . . . The breaker has gone out before them, they have broken through, they pass through the gate, they go out through it, and their king passes on before them, Jehovah at their head!" (ASV).

Star in Perseus:
    Atik: Who Breaks

Note: There is one interesting fact I must make note of. Remember that two stars in Aries are Al Nath, "The Wounded," and Al Sheratan, "The Bruised." The place of Roman crucifixion always took place on an elevated location near well-traveled public roads. This was done to make a public spectacle of the one being executed. Although the precise place of Jesus's crucifixion is under debate, the most widely accepted place of Calvary is situated (by latitude and longitude) so that at the time the sun went dark that afternoon, the cross would have been directly under El Nath and Al Sheritan to anyone looking up at a thirty-five-degree angle, looking southwestward, from the road below.

I believe this to be more than an interesting bit of trivia. God, in His infinite wisdom and power, set all this in place, at creation, to come true four thousand years later. God is awesome.

Christmas (birth of Jesus): Occurred on September 29, 2 BC. The sun was in Virgo (in conjunction with Spica: "The Branch").

Note: There seems to have always been a lot of speculation as to the real date (including the year) of Jesus's birth. The idea of King Herod's death in secular records predating the supposed date of Jesus's birth, or that shepherds would not have been in the fields with the sheep in September, may or may not have any merit. I believe one thing does have merit for dating this event. And that is the zodiac. So let's continue and you can judge.

Mars (The Wounded) was in Virgo (The Seed) in conjunction with Porrima (The Coming Prophet). This is the coming forth of the Prophet.

Jupiter (The Lord Has Set Free) was in Virgo (The Seed). The One who would set free was coming forth.

Venus (The Beloved, The Awaited One) was in Leo (The Lion Ruling). The awaited One was coming to rule (spiritually).

Mercury (Going and Returning Again) was in Libra (The Required Price Paid) in conjunction with the star Zuben Al Genubi (The Price Which Is Deficient) and the moon (The Lord's Glory). The One who would go and return again, that is, the Lord of glory, came to pay the required price because man could not pay it, although he was constantly trying.

Saturn (Resting, The Work Completed) was directly on the boundary (just above the ecliptic) between Gemini (The Two Comings) and Orion (The Brightness of His Coming). His coming, His sacrifice, would complete the work that no other man could do (the required price).

Uranus (Ruling, Fearful Judgment) was in The Band (That Which Joins the Former and Latter Church) closest to the "latter" fish. He came to rule over the church, yes, even with a certain fearful judgment.

On the spring equinox, the sun was in Pisces
On the summer solstice, the sun was in Cancer.
On the fall equinox, the sun was in Libra.
On the winter solstice, the sun was in Capricorn.

Virgo: The Seed of the Woman
    Isaiah 7:14
    Therefore the Lord Himself will give you a sign: Behold, the virgin shall conceive and bear a Son, and shall call His name Immanuel. (KJV).

Stars in Virgo:
>   Spica (located in the wheat ear) (the seed carried by the virgin): The Branch
>   Vindemiatrix: the Branch Who Comes

Virgo's decans:
>   Coma Berenices: The Desired of Nations
>   Centaurus: The Despised (the man of double nature in humiliation)
>   Bootes: The Coming (the exalted shepherd and harvester)

Coma Berenices: The Desired of Nations
>   Haggai 2:6–7
>   For thus says the LORD of hosts: Once again, in a little while, I will shake the heavens and the earth and the sea and the dry land; and I will shake all the nations, so that the desire of all nations shall come, and I will fill this house with splendor, says the LORD of hosts. (NRSV).

Star in Coma:
>   Al Zimach: The branch, the seed

Centaurus: The Despised (The man of double nature in humiliation)
>   Isaiah 53:3
>   He was despised and rejected—a man of sorrows, acquainted with bitterest grief. We turned our backs on him and looked the other way when he went by. He was despised, and we did not care. (NLT).

Star in Centaurus:
>   Toliman (Alpha Centauri): The Heretofore and the Hereafter

Bootes: The Coming (The exalted shepherd and harvester)
>   Revelation 14:15–16
>   Another angel came out of the temple, calling with a loud voice to the one who sat on the cloud, "Use your sickle and reap, for the hour to reap has come, because the harvest of the earth is

fully ripe." So the one who sat on the cloud swung his sickle over the earth, and the earth was reaped. (NRSV).

Star in Bootes:
    Arcturus: He Comes

Note: I will here give the reasons why I believe September 29, 2 BC, is the correct date for the birth of Jesus. I'm sure many people are asking how it could be 2 BC when from 2 BC to AD 33 there would be thirty-five years. Our calendar contains a two-year error. When historians originally attempted to count backward to the birth of Christ, they did so by accounting for the reign of kings. One of the kings counted was actually ruler of his country twice, the second time under a different name for a period of two years. When the years of rule were counted, this king's reign was mistakenly counted twice.

First, Virgo represents the Seed of the Woman. And the star located in her "belly" is Porrima. The "star" is actually a double star. The companion star has an orbital period of 168+ years. This is how I interpret Porrima. The main star in the "belly" of Virgo is the virgin. The companion star is the prophet (Jesus) that Moses spoke of in Deuteronomy 8:15. From its last point of maximum separation (from earth's viewpoint), counting backward, the companion star would have been "separating" from the main star in the year 2 BC (when Jesus was "separating from" or coming forth from Mary).

Secondly, on August 8 of 3 BC, Venus came into conjunction (close proximity) with Jupiter near Regulus (the brightest star in Leo). The moon (The Lord's Glory) occulted (came directly in front of) Asellus Borealis (the Northern Donkey Colt) in Cancer near the star cluster Praesepe (The Manger). On August 12, Jupiter was occulted by Venus. On August 16, Venus moved from in front of Jupiter and was about to occult (pass in front of) Regulus, which occurred on August 17. Venus moved from in front of Regulus on August 19 and continued moving away from Jupiter and Regulus while Jupiter, less than one degree from Regulus, continued to move even closer to Regulus. On September 4, the moon again passed between Asellus Borealis and the Praesepe, this time a little farther south toward Praesepe. On September 6, Regulus was occulted by Jupiter.

Here I need to say that Regulus has been seen as the *king* star and Jupiter as the *king's* planet from antiquity. Venus has always been viewed as representing the "Beloved" or "The Awaited One." Also, it's important to note that only planets are referred to as morning or evening "stars." And in the Greek New Testament, the word translated as *star* means "spread," as the stars are "spread out" across the universe. This conjunction lasted for twenty days. Then on November 8, still moving away from Regulus, Jupiter would have appeared to slow in its motion until November 16 when its motion stopped and stood still until December 12, 3 BC, when it began moving back toward Regulus.

The reason for this "standing still" and reversing was due to its relative position and speed in its orbit in relation to earth. (See the diagram and explanation below.) It then would have again been close enough in conjunction with Regulus to appear as one. During this time, on November 25, the moon again directly passed through Praesepe. In December, the moon again passed between Asellus Borealis and the Praesepe. Again on January 9, 2 BC, the moon passed through the Praesepe. Again, on February 11, 2 BC, Regulus was occulted by Jupiter, while on February 15, the moon passed through the Praesepe. Regulus's occultation by Jupiter lasted until March 2. On March 14, the moon passed through the Praesepe. Meanwhile, Jupiter was moving away from Regulus only about one degree, and then on March 27 Jupiter stopped, reversed, and started moving back over Regulus again to occult Regulus on April 28.

On May 4, Venus was less than three degrees from the star Wasat (The Appointed) in Pollux (the first coming) of Gemini. On May 8, the moon passed through the Praesepe. As Jupiter moved away from Regulus, revealing it on May 17, Venus moved through the Praesepe. On May 22, Mars moved to within two degrees of Wasat in Pollux of Gemini. On June 1, Mercury moved to within two degrees of Wasat in Pollux of Gemini. On June 11, Mercury passed through the Praesepe while Venus nearly occulted Regulus, then did occult Jupiter on June 17 and remained in conjunction with Jupiter until late June (2 BC). On June 20, Mars passed through the Praesepe. On July 29, the moon passed between Asellus Australis and the Praesepe. On August 25, the moon passed between Asellus Australis and the Praesepe. On September 21, the moon passed through the Praesepe.

The next question is this: when did the Magi actually arrive to see the new king? About the only thing we know for sure is that when they arrived, Jesus's family was in a house and not in a stable. Obviously, they arrived after Mary and Joseph had moved from the stable to a house. There has been much conjecture over how long their trip was. I have heard theories ranging from one to two years. Does the zodiac support any date? I believe it does.

From late June (2 BC) until August, Jupiter would keep making its trek westward until, on August 27, Jupiter, Venus, Mercury, and Mars were in conjunction, less than two degrees from one another in Leo and headed toward Virgo. The sun had just entered into Virgo, and the moon (only two days from a "new moon") was occulting Regulus. Jupiter, our main character, would continue its movement westward into Virgo. On December 25, 2 BC, Jupiter again stopped and remained stationary for six days. The Magi were then in Jerusalem inquiring about the newborn King. And on December 25, at the ordinary time for the Magi's predawn observations, Jupiter would have been seen in meridian position (directly over Bethlehem) at an elevation of sixty-eight degrees above the southern horizon. This precise position would show the planet shining directly down on Bethlehem while it was stationary among the stars. With Bethlehem being only six miles south of Jerusalem, the Magi would have no trouble getting there that day.

Thirdly, the positions of the planets as listed above on September 29 are irrefutable.

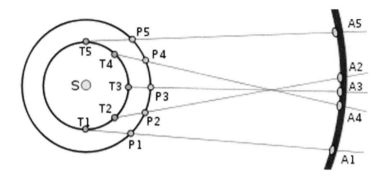

Credit: Rursus, Creative Commons Attribution 3.0 Unported.

Jupiter becomes "stationary" at its times for retrogression and progression. When we look at Jupiter, we see the planet normally moving eastward each evening through the fixed stars. This apparent movement is called "proper motion." The earth, however, is moving in its orbit around the sun faster than that of Jupiter. When the earth reaches point T1, an observer would see Jupiter nearly along the same line as the earth's own orbital movement. When the earth is traveling more or less in a direct line toward Jupiter, the planet will continue to show "proper motion." But when earth reaches position T2, it is no longer heading toward Jupiter. The faster velocity of the earth, as it makes its turn to T2 and beyond, causes the apparent motion of Jupiter to slow down. This continues until the earth reaches T3.

At that point, the speed of the earth in relation to Jupiter is the same as Jupiter's. That is when Jupiter appears to become stationary within the background of the fixed stars. As the earth progresses from T3 to T4, it has greater relative speed than Jupiter, and this causes Jupiter to retrogress. The planet reverses its motion and travels westward through the stars. At T4, however, the speed of the earth and Jupiter are again matched (relative to each other) and Jupiter stops its reverse motion. When T4 is passed, Jupiter returns to "proper motion." This is what happened when Jupiter came in contact with the star Regulus on three different occasions in the late part of 3 BC and the early part of 2 BC and again in late 2 BC (December).

The Exodus (First Passover): Occurred on March 28, 1453 BC. The sun was in Aries.

    Jupiter (The Lord Has Set Free) was in Aries (The Ram Sacrificed) in conjunction with the sun (The Father's Glory). The Father was in control. Through the Passover sacrifice, the Lord was about to set Israel free.
    Mercury (Going and Returning Again) was in Aries (The Ram Sacrificed). The sacrifice was the release of the Lord.
    Saturn (Resting, The Work Completed) was in Aquarius (The Blessing Poured Out). The work was complete; the last "plague" on Egypt brought about the blessing of release and ended bondage.
    Mars (Wounded) was in Capricorn (Life out of Death) in conjunction with the star Deneb Al Gedi (Cut Off). The captive (wounded) Hebrews (through which the Messiah would come) were being given life (freedom) from death (slavery). They would be able to worship and serve God as He wanted.
    Venus (The Beloved, The Awaited One) was in Aquarius (The Blessing Poured Out). The Lord was bringing freedom to the Hebrews.
    Uranus (Ruling, Fearful Judgment) was in Sagittarius (Going Forth). The Lord had fearfully ruled over Egypt and the Hebrews were going forth.
    The moon (The Lord's Glory) was in Taurus (The Judge Coming to Rule). The Lord had gloriously judged Egypt.

    On the spring equinox, the sun was in Aries.
    On the summer solstice, the sun was in Cancer.
    On the fall equinox, the sun was in Virgo.
    On the winter solstice, the sun was in Capricorn.

Note: Here I am marking Passover instead of the actual exodus, which occurred after the Passover night. After all, Passover was the "main event."

The Exodus (Passover) is marked by the same constellation as the crucifixion. The only difference is that the sun was not as far into Aries at this time and, due to precession, the constellations the sun was in on the equinoxes and solstices were not all the same. Remember the Jewish calendar was changed at the time of the exodus to reflect their New Year.

Aries: Glory out of Humiliation
> Revelation 5:12
> saying with a loud voice: Worthy is the Lamb who was slain to receive power and riches and wisdom, and strength and honor and glory and blessing! (NKJV).

Stars in Aries and their meaning:
> Al Botein: The Treading under Foot
> Al Thuraiya: The Multitude, The Abundance

Aries's decans:
> Cassiopeia: Freed and Preparing for Her Marriage to the Redeemer
> Cetus: The Great Enemy Subdued and Bound
> Perseus: The Breaker Delivering the Redeemed

The following event dates correspond to our present-day calendar, not the Creation Calendar, because computer models depicting the position of the sun on the ecliptic do not account for the differences in calendars.

The Flood: Noah entered the ark on October 30, 2310 BC. The sun was in Ophiuchus (Scorpio/Ophiuchus share the ecliptic. Although Ophiuchus is not considered to be on the ecliptic by modern astronomy, I refer to it as such for biblical astronomy).

The sun (The Father's Glory), Mercury (Going and Returning Again), and Venus (The Beloved, The Awaited One) were in conjunction right at the borderline between Scorpio (Wounded) and Ophiuchus (The Serpent Held). Venus (The Beloved, The Awaited One) was on the ecliptic, straddling the north/south boundary line, Mercury (Going and Returning Again) was just north of the line in Ophiuchus, and the sun was in Ophiuchus. The battle for man between God and Satan had gone to Satan until now. (There are only eight righteous left.) The sun in Ophiuchus represents the turn the battle is about to take. Mercury and Venus are over Antares (The Wounding) (the heart of Scorpio). This means that Scorpio is in trouble. He is about to lose.

Mars (Wounded) was in Virgo (The Seed). The Lord (the wounded one) was coming forth.

Jupiter (The Lord Has Set Free) and Saturn (Resting) were in conjunction in Capricorn (Life out of Death). The Lord was about to set man free by destroying the wicked. Rest would come to those who are righteous. Out of the death of the unrighteous, life would come; is it a coincidence that the name Noah means "He Will Give Us Rest"?

Uranus (Ruling, Fearful Judgment) was in Libra (The Required Price Paid) near the star Zuben Al Genubi (The Price Which Is Deficient). The ruling Lord was passing fearful judgment on man because what man has done (animal sacrifice and "good" living) cannot satisfy God's required price.

The moon (The Lord's Glory) was in Taurus (The Judge Coming to Rule). The Lord was about to bring about harsh judgment.

On the spring equinox, the sun was in Aries.
On the summer solstice, the sun was in Cancer.
On the fall equinox, the sun was in Libra.
On the winter solstice, the sun was in Capricorn.

The rain began seven days later (November 6). Venus and Mercury had changed places. The moon was in Leo (His Enemies Destroyed). The other planets had not changed constellations.

One year later, the earth became dry, Noah built an altar of sacrifice, and God made a covenant with Noah on November 17, 2309 BC.

    The sun (The Father's Glory) was in Sagittarius (Going Forth).

    Venus (The Beloved) was in conjunction with Uranus (Ruling) in Libra (The Required Price Paid).

    Mercury (Going and Returning Again) was in Ophiuchus (The Serpent Held).

    Mars (Wounded) was in conjunction with Jupiter (The Lord Has Set Free) in Aquarius (The Blessing Poured Out).

    Saturn (Resting) was in Capricorn (Life out of Death).

    The moon (The Lord's Glory) was in Gemini (The Two Comings).

Creation: Occurred October 14–20 (seven days), 3996 BC. The sun was in Ophiuchus.

Day 1: There was no sun or moon. There were no stars or planets.

Day 4: The sun (The Father's Glory) was in Ophiuchus (The Serpent Held) in conjunction with Mercury (The Going and Returning Again) near the star Saiph (Bruised). The stage was set already as God spread the stars and planets out in space. The story was already told. Even before man, the Lord (the Bruised) was already defeating Satan. This was prophecy declared by God. It will happen.

Saturn (Resting) was in Libra (The Required Price Paid). God would provide rest by the price of blood to cover sin when man falls.

Jupiter (The Lord Has Set Free) and Venus (The Beloved, The Awaited One) were in Sagittarius (Going Forth) with Venus passing into Capricorn (Life out of Death) that night. The Lord, who would set free, the Beloved, would come forth and bring spiritual life out of spiritual death.

Mars (Resting) was in Aquarius (The Blessing Poured Out). The Lord would bring rest in the blessing He would pour out.

Uranus (Ruling, Fearful Judgment) was in Virgo (The Seed). The Messiah would come to rule.

The moon (The Lord's Glory) was in Leo (His Enemies Destroyed). The glorious Lord would destroy His enemies.

Day 8: The sun (The Father's Glory) was in Opheuchus (The Serpent Held) directly between Scorpio's (Wounding) sting and Ophiuchus's foot. The wounding of all mankind has happened.

Mercury (The Going and Returning Again) was in conjunction with Jupiter (The Lord Has Set Free) in Sagittarius (Going Forth). The promised One, who will set free, go, and return, will come forth.

Venus (The Beloved, The Awaited One) was in Capricorn (Life out of Death). The Lord will bring life out of death.

Mars (Wounded) was in Aquarius (Blessing Poured Out). The wounded One will pour out blessings.

Uranus (Ruling, Fearful Judgment) was in Virgo (The Seed). The One who will rule in fearful judgment will come forth.

Saturn (Resting) was in Libra (The Required Price Paid). He will pay the required price and bring rest.

The moon (The Lord's Glory) was in Virgo (The Seed) in conjunction with the star Spica (The Branch). He, the Lord, is the Messiah to come forth.

There was no spring equinox, no summer solstice, and no fall equinox. On the winter solstice, the sun was in Capricorn.

Scorpio: Wounding

> Isaiah 53:5
>
> But He was wounded for our transgressions, He was bruised for our iniquities; the chastisement for our peace was upon Him, and by His stripes we are healed. (NKJV).

Stars in Scorpio:
> Antares: The Wounding
> Leseth: The Perverse (in Scorpio's tail)

Scorpio's decans:
> Serpens: The Conflict for the Crown
> Ophiuchus: The Evil One Held
> Hercules: The Mighty Victor

Serpens: The Conflict for the Crown

> Revelation 12:7–9
>
> Then there was war in heaven. Michael and the angels under his command fought the dragon and his angels. And the dragon lost the battle and was forced out of heaven. This great dragon—the ancient serpent called the Devil, or Satan, the one deceiving the whole world—was thrown down to the earth with all his angels. (NLT).

Stars in Serpens:
> Unuk: Encompassing
> Cheleb: The Serpent Enfolding (in the serpent's jaw)

Ophiuchus: The Evil One Held

> 2 Thessalonians 2:5–7
>
> Do you not remember that I told you these things when I was still with you? And you know what is now restraining him, so that he may be revealed when his time comes. For the mystery of

lawlessness is already at work, but only until the one who now restrains it is removed. (NRSV).

Star in Ophiuchus:
    Triophas: Treading under Foot

Hercules: The Mighty Victor
    2 Thessalonians 2:8
    And then the lawless one will be revealed, whom the Lord Jesus will destroy with the breath of his mouth, annihilating him by the manifestation of his coming. (NRSV).

Star in Hercules:
    Guiam: Punishing

Note: I *must* make some interesting notes here.

By my calculations and by most other authorities, October (on our present Gregorian calendar) is the actual month of creation. October corresponds to the first month (Tishri) of the Hebrew calendar until the Exodus when God changed the first month to Nisan (our late March/early April). Actually, although the time frame would correspond to October, there are no corresponding dates on the Hebrew calendar. The last day in Tishri (the first month) corresponds to October 13. The first day in Heshvan (the second month) corresponds to October 21. Creation week (seven days) simply sets in between those dates. We know that God created all that is in the physical universe in six days (twenty-four-hour days) and then rested from creating on the seventh day. In the month of October, as the sun passes through the constellations on its yearly trek around the ecliptic, it stays in Scorpio only thirteen days.

Here is how I believe the first eight days of world history occurred (excluding the creation of angels, which occurred before man):

Day one: God creates "earth" and "heavens"; God causes light, separating light from darkness.

Day two: God makes an expanse (space) to separate the water (above space) from the water (below space).

Day three: God makes dry land appear; God has the earth bring forth vegetation on the dry land.

Day four: God makes the sun, moon, and stars. The zodiac is set in place with the sun being set in the constellation Scorpio. The moon and planets are set in their places and paths to be in place at appointed times for earth's historical events.

Day five: God brings forth life in the sea (fish) and in the air (birds).

Day six: God brings forth land animals (including insects); God makes mankind.

Day seven: God rests from creating.

Day eight: The Hebrew narrative indicates that Satan came in the morning, deceived Eve and Adam into sinning, and went away smirking and thinking, *Gotcha*. Adam and Eve made their makeshift Fruit of the Looms and warily went about their business. When they "heard" God in the garden that evening, they hid. That same evening, Adam and Eve played the blame game with God. Remember that Adam said, "The woman you gave me," and Eve said, "The devil made me do it." So then God curses Satan and tells him,

> Because you have done this, you are cursed above all cattle, and above every beast of the field; you will slither upon your belly, and you shall eat dust all the days of your life and I will put enmity (hatred) between you and the woman, and between your offspring and her offspring; hers shall bruise your head, and you shall bruise his heel" (Genesis 3:14–15 TMNT paraphrased).

Adam and Eve are put out of the garden.

Why do I believe it happened that way?

1) On day four of creation, the sun, moon, stars, and all "space matter" were placed by God right where it all needed to be. The sun was in Scorpio (representing Satan wounding the Christ) but would be

in Scorpio only four more days. On October 21, the first day after creation week (day eight), the sun was directly on a line between Scorpio's tail (sting) and Ophiuchus's foot. I believe this is the day that Satan tricked Eve and Adam into sinning. The gospel story in the stars bears this out.

2) The closest star to being a true "pole star" was Thuban, the brightest star in the constellation Draco (representing Satan). Thuban means, "The Subtle." Genesis tells us that the Serpent was more subtle than any other creature.

3) Creation week parallels the Passover celebration on the Exodus Calendar. Passover is the lamb (of God) sacrificed (bruised, not killed, by Satan) by shedding its (His) blood, as God had to do to clothe Adam and Eve with animal skins for their sin.

4) The planets were in the "right" corresponding constellations to support (actually prove) the history.

5) I believe Satan would not have wasted any time getting to Eve. Why would he allow them time to enjoy "paradise"? Why would he allow them time to get used to God's presence and thereby become harder to deceive?

Let's go on to the Age of Aquarius. I saved this for last because it is a pivotal date in history. Here I will only give the date that we entered the new age, but I will have more to say about it later.

We entered into the Age of Aquarius on March 20, 2004, at 6:47 a.m. (the vernal equinox).

The Sun (The Father's Glory) and the Moon (The Lord's Glory) were in Aquarius (The Blessing Poured Out). During this age, God will pour out again, in a mighty way, His Spirit on all of His believers.

Jupiter (The Lord Sets Free) was in Leo (The Ruling King). He will set free and rule during this age

Saturn (Resting) was in Gemini (His Rule on Earth). The Lord's people are resting spiritually awaiting His return.

Mars (Wounded) was in Taurus (The Judge Coming to Rule). The "wounded" Lord will come again in judgment to rule.

Venus (The Beloved, The Awaited One) was in Aries (Glory Out of Humiliation). The Lord will come in glory in this age, not humility.

Mercury (The Going and Returning Again) was in Pisces (Deliverance Out of Bondage). The returning Lord will deliver His people out of bondage.

Uranus (Ruling, Fearful Judgment) was in Aquarius (The Blessing Poured Out). The Lord will return in judgment of the wicked but to bring blessing to His children.

CHAPTER 10

The Hebrew Cycle and History

So far I have shown how the stars tell the whole gospel story past, present, and future. And I have shown how the precession of the equinoxes verifies a timeline of critical events in earth's past history from creation up to AD 2004. Now is there anything else which will add validity to our timeline? Yes, there is.

I have focused on the sun traveling along the ecliptic in years (earth's solar years). But there are other kinds of years and cycles. Technically, a "year" is the time it takes for "something" to make one revolution around "something else." So let's see what other kinds of astronomical years there may be.

1) Lunar year: The time it takes for the moon to go through twelve full cycles. The lunar year is equal to 154.37 days.
2) Great year: The time it takes for the sun to make one complete cycle around the ecliptic. The great year is equal to about 25,700 earth years.
3) Galactic year: The time it takes for our solar system to complete one cycle around the galaxy's center. The galactic year is equal to approximately 230 million earth years.
4) Grand year: A Hebrew cycle of 600 years defining the solar year.

So, if the grand year is equal to the solar year, and the solar year is what I have been using to explain our timeline, why do I list it separately? I list it separately because the grand year was discovered and used from the earliest ancient times. Giovanni Cassini (1625–1712) found that the period of six hundred years, of which we find no intimation in any records but those of the Jewish nation, referred to by Josephus, and called the grand year, is one of the finest ever invented. It brings out the solar year more exactly than that of Hipparchus and Ptolemy, and the lunar month within one second of what is determined by modern astronomy.

THE FUTURE IS IN THE STARS

Cassini also urged that nothing but the observations of those who lived to see celestial bodies return to the same places could have originated this wonderful period.

This argument, again brought forward by Sir W. Drummond, appears to have had weight with him in his conversion to a reverential belief in the authority of those Holy Scriptures he had once undervalued. Cassini, in verifying this ancient calculation, had the use of observations made by means of the instruments of then modern science, and from these he could ascertain what the patriarchs might know by visual inspection of the course of the heavenly bodies. In their lives, one man might observe twenty or thirty revolutions of Saturn, sixty or eighty of Jupiter, and many more of the terrestrial planets. By this great cycle of six hundred years, the lunar month is reckoned at 29 days, 12 hours, 44 minutes, 3 seconds; the solar year at 365 days, 5 hours, 51 minutes, 36 seconds: not that this division was so made in the ancient tradition, as made by computation, but is the result given of the actual completion of the cycle, which might be observed by those whose lives were of sufficient length. After the first completion of the first six hundred years had been witnessed, every succeeding year would furnish another, a new proof of its accuracy.

Is it a coincidence then that from creation to the Age of Aquarius, which I will show is very significant in the zodiac timeline, is an exact multiple of six hundred years? And is it a coincidence that the great year is an exact multiple of six hundred?

CHAPTER 11

Earth's Magnetic Field Decay

Up to this point, I have focused our attention on the zodiac and supporting sources only. Now I believe that no other source is necessary, but, for the skeptical, I will include some other information.

One more thing to consider as far as dating the age of the earth (a creation date) is the earth's magnetic field. The earth's magnetic field is getting weaker; it is decaying. It is a scientific fact. Dr. Russell Humphreys gives us a theory based on scientific evidence. In his paper "The Earth's Magnetic Field Is Young," Dr. Humphreys says the earth's magnetic field has decreased by about 14 percent since 1829. Subsequent measurements indicate that the intensity of the earth's magnetic field was about 40 percent greater in AD 1000. The data show that the field intensity at the earth's surface fluctuated wildly up and down during the third millennium before Christ (the time of the flood). Paleomagnetic data show that while the geologic strata were being laid down, the earth's magnetic field reversed its direction hundreds of times. This "dynamic decay" theory contradicts evolutionary theories and shows that with a significant loss of energy during the Genesis flood, the age of the field would be about 6,000 years.[1] This fits in well with the face-value biblical age of 6,000 years.

It also fits the biblical astronomy age of 6,000 years for the earth.

The earth is not alone. In 1974 and 1975, the Mariner 10 spacecraft measured Mercury's magnetic field strength with its onboard magnetometer and sent the data to earth. In 2008, Messenger flew past Mercury and captured a magnetic field measurement. Unsurprisingly, Mercury's magnetic field strength had diminished since 1974. In fact, when compared with Humphrey's creation model, the numbers aligned perfectly, indicating a solar system age of 6,000 years.

1. Russell Humphreys. "The Earth's Magnetic Field Is Young." Institute for Creation Research, 1993.

CHAPTER 12

Hebrew Feasts and Their Relation to Church Events

Remember that I said the zodiac is world history past, present, and future. Until now I have talked only about the past, but what of the future? Before checking the zodiac for future dates, we need to know what to look for. Let's start with some general ideas.

God instituted feasts (holy days) for the Jews to observe. These feasts correspond with history and Christian events. Those feasts and their corresponding events are as follows:

Passover: Yeshua's sacrifice
Unleavened bread: Yeshua, the sinless one
Firstfruits: Yeshua's resurrection
Pentecost: Falling (Giving) of the Holy Spirit

These first four have already been fulfilled.

Rosh Hashanah (Trumpets): His second coming foretold (announced by the sounding of the trumpet)
Yom Kippur (Day of Atonement): Yeshua's second coming
Booths (Tabernacles): Yeshua with us

These three are yet to be fulfilled.

What do these feasts have to do with the zodiac and the gospel story and our timeline? Just this: there are still three feasts to be fulfilled. Those three all relate to the second coming of the Christ (the announcement of His coming, His coming, and His presence with us). Again, the zodiac is world history past, present, and *future*. Regarding the future, does the zodiac tell us the time of the second coming? Matthew 24:36 tells us that no man knows the day or the hour (and I agree with that verse). But I also know that in Matthew 24:32–33, Jesus said, "Now learn a parable of the fig tree;

When his branch is yet tender, and puts forth leaves, you know that summer is near and so likewise, when you shall see all these things, know that it is near, even at the doors." (KJV). And Matthew 24:29–30 says, "Immediately after the tribulation of those days shall the sun be darkened, and the moon shall not give her light, and the stars shall fall from heaven, and the powers of the heavens shall be shaken. And then shall appear the sign of the Son of man in heaven: and then shall all the tribes of the earth mourn, and they shall see the Son of man coming in the clouds of heaven with power and great glory." (KJV).

Let's take a look at these verses:
Verse 36: No one—not the angels, not even Jesus—but God knows the *exact* time of the second coming.
Verses 32–33: Jesus says there shall be "signs" that will precede His return. When these signs are seen, His return is near. The phrase "even at the doors" means "imminent" or "about to enter."

Verses 29–30: What do a darkened sun and moon and falling stars sound like? The reference to the sun and moon sounds like a solar and lunar eclipse, respectively, to me. And Revelation 6:12 adds that the moon will turn blood red, the color the moon appears during an eclipse. And the falling stars would be meteors during a meteor shower. (Many people even use the term "shooting star.")

Not that I will try to predict any specific time for the second coming of Christ (many have done that in the past, and all have failed), but as a theoretical possibility and something to just think about, allow me to make this suggestion: we know that certain signs in the zodiac represent the second coming of Christ. There is more than one sign, along with its decans, that speak of the time of the second coming. But only one speaks specifically of God "pouring out His blessing" on the church. That sign is Aquarius.

The prophet Joel says that God will "pour out His Spirit . . . in the last days." I believe this is represented by Aquarius. All of the decans associated with Aquarius represent the second coming. Piscis Austrinus is the "latter" church receiving the "blessing" in the last days. Pegasus is the soon return. Cygnus is the sure return. Can it be that God has given us this "sign" to

THE FUTURE IS IN THE STARS

give us His timeline (in general) for the second coming? Also, what of the eclipses and meteor showers or of the three remaining Jewish feasts? Since we have the "formula" to determine how long the zodiac signs last, it's easy enough to see that the Age of Aquarius started in AD 2004 and will go on until about the year AD 4578. This allows a fair amount of time in the Age of Aquarius for the second coming to occur (if we believe that things can get much worse).

As I said before, there are overlaps of signs with Aquarius. Pisces overlaps on the early date side and Capricorn overlaps on the later date side. Could the second coming be, in the Pisces-Aquarius overlap, symbolizing His coming during the last stages of the church age (the latter fish represents the latter church) or in the Aquarius-Capricorn overlap symbolizing His coming when all seems so bleak (dead) and He brings life out of death? Since the first four Jewish feasts were fulfilled in the months (on the days of the months) when they occurred, is there any reason not to believe that the last three feasts will occur the same way? The last three feasts occur in September to early October based on new moons.

Solar and Lunar eclipses can occur at any time of the year. Meteor "showers" occur at the same time of the year on an annual basis. One of the busiest times for meteor showers is September and October. In September, showers radiating from Aries (the Ram) and Taurus (the Judge) easily reach twenty per hour. Also in October is the Orion (The Brightness of His Coming) shower. There is also some activity coming out of Leo (the Rending Lion) in October.

Is this coincidence or divine providence?

Also, don't forget that Thuban ("Subtle") was the pole star at creation. Polaris (Revolved Around), in Ursa Minor (The Stronghold of the Saved), is now the pole star and will be for about a thousand plus years. Who does the whole history of the world really revolve around?

Again, I want to emphasize that I do not claim to know or have discovered the exact time of the second coming of Christ. But I do believe that God has given us the signs of the zodiac and "the signs of the times" to determine when the time is near.

According to astrology, the Age of Aquarius is an age of spiritual awakening of the inner self in relation to the cosmos and the forces "out there." It is supposed to be an age of tremendously increased knowledge and understanding and inward peace, which will bring about peace among all men. This "awakening" age could possibly bring about the next evolutionary stage of mankind, transforming him into some "higher form" of spiritual being. Personally, I think they have been watching too much *Star Trek*. There are many problems with this thinking. We need to deal with only one. Even among astrologers, they can't agree on the right date for entering into the Age of Aquarius. Dates range from AD 2000 to the IAU's estimated date of AD 2579.

There is one more occurrence to present as far as our zodiac timeline is concerned. During the Age of Aquarius, there will be an astronomical event that happened only once before in earth's history. Beginning September 7, 1954 BC, all six planets were above the horizon at the same time. They moved generally closer together until November 9, when Venus moved westward past Saturn. Around November 23, Venus began to slow and move closer to the ecliptic, which it crossed on December 13. After crossing the ecliptic, Venus begins moving back toward the other planets. On January 10, 1953 BC, all six planets were above the horizon on the eastern side of the meridian. On January 27, there was a solar eclipse. The planets then began to move farther from one another. What's the significance of this? 1953 BC is the year that God made His covenant with Abraham. I have not been able to ascertain the exact date of this biblical (and historical) event of that year, but I would venture to say that it was around January 27 (of our calendar). And it was a new moon so Abraham would have had a great view of the stars when God told him to go count the stars.

The sun was in Aquarius (the Blessing Poured Out).
Mars, Saturn, and Uranus were in Capricorn (Life out of Death). Hebrews 11:12 reads,

> Therefore sprang there even of one, and him **as good as dead**, so many as the stars of the sky in multitude, and as the sand which is by the sea shore innumerable (KJV)(Emphasis mine).

Judging by the importance of that event, I'm convinced that when the next conjunction of the same six planets occurs, it will mark a very important event. When will this happen? It will occur from May to September of the year AD 2040. From around May 12, all six planets will be above the horizon at the same time. They will continue moving closer to one another until on September 8, all except Uranus, and will be within fifteen degrees of one another.

First, let me set the stage for the "main event." On August 22, the sun will be in Leo right on top of Regulus. Mercury and Venus will also be in Leo. Jupiter, Mars, and Saturn will be in Virgo. The moon will be in Aquarius and Uranus will be in Cancer. I'll leave it to you to remember the meanings of the planets and constellations.

On September 5, all five planets will be in conjunction. The sun will still be in Leo and the moon (one day before new moon) will have moved into Leo near Regulus. Uranus will be still in Cancer.
Let's now move on to the main date: AD September 8, 2040.

The sun will be in Leo.
The five planets—Mercury, Venus, Mars, Jupiter, and Saturn (all representing some aspect of the Lord's glory)—and the moon (the Lord's glory) will be in Virgo and congregating around Porrima. Remember that Porrima represents the "prophet" Jesus coming forth. At this time, the companion star of Porrima (the prophet) will again be moving away (at its farthest point) from the main star. (All five planets will be moving toward Spica, The Seed).
Uranus will be in Cancer.

On September 21, all five planets will still be in conjunction in Virgo.
Jupiter and Saturn will still be near Porrima.
Mars and Venus will be near Spica, and Mercury will be directly on Spica.
Uranus will be in Cancer (still above the horizon).
The moon will be in Cetus (the enemy subdued and bound) just below the ecliptic.

The five planets and the sun will remain in Virgo through September 27.
Leading up to this period, there are a number of solar and lunar eclipses:

November 30, 2039: Lunar eclipse, the sun will be in Scorpio near Graffias and the moon will be in Taurus.
December 15, 2039: Solar eclipse, the sun and the moon will be in Scorpio near Antares and in conjunction with Mercury (Going and Returning Again).
May 11, 2040: Solar eclipse with the sun and moon in Aries in conjunction with Jupiter.
May 26, 2040: Lunar eclipse, the sun will be in Taurus in conjunction with Jupiter and the moon in Scorpio in conjunction with Graffias.

If that's not enough, the remaining three Hebrew feasts:
Rosh Hashanah (Trumpets): His second coming foretold (announced by the sounding of the trumpet)
Yom Kippur (Day of Atonement): Yeshua's second coming
Booths (Tabernacles): Yeshua with us
begin on September 8 and end on September 27.

Let's consider something Jesus said in Matthew 25:1–13.

> Then shall the kingdom of heaven be likened unto ten virgins, which took their lamps, and went forth to meet the bridegroom. And five of them were wise, and five were foolish. They that were foolish took their lamps, and took no oil with them: But the wise took oil in their vessels with their lamps. While the bridegroom tarried, they all slumbered and slept. And at midnight there was a cry made, Behold, the bridegroom cometh; go ye out to meet him. Then all those virgins arose, and trimmed their lamps. And the foolish said unto the wise, Give us of your oil; for our lamps are gone out. But the wise answered, saying, Not so; lest there be not enough for us and you: but go ye rather to them that sell, and buy for yourselves. And while they went to buy, the bridegroom came; and they that were ready went in with him to the marriage: and the door was shut. Afterward came also the other virgins, saying, Lord, Lord, open to us. But he answered and said, Verily I say unto you, I know you not. Watch therefore, for ye know neither the day nor the hour wherein the Son of man cometh. (KJV).

Could there be another way to interpret the five planets in Virgo? Could the five planets represent the five wise virgins who were prepared for the bridegroom at His coming?

Again, I want to emphasize that I do not claim to know or have discovered the exact time of the second coming of Christ. But I do believe that God has given us the signs of the zodiac and "the signs of the times" to determine when the time is near. Could this astronomical event represent the second coming? *Something* very important is on God's timetable.

CHAPTER 13

The Great Pyramid of Giza

Remember when I spoke of the history of astronomy and that Josephus spoke of a "pillar of stone and a pillar of brick"? Here again is his writing:

> They also were the inventors of that peculiar sort of wisdom which is concerned with the heavenly bodies, and their order. And that their inventions might not be lost before they were sufficiently known, upon Adam's prediction that the world was to be destroyed at one time by the force of fire, and at another time by the violence and quantity of water, they made two pillars, the one of brick, the other of stone: they inscribed their discoveries on them both, that in case the pillar of brick should be destroyed by the flood, the pillar of stone might remain, and exhibit those discoveries to mankind; and also inform them that there was another pillar of brick erected by them. Now this remains in the land of Siriad to this day.[1]

Ken Johnson, PhD, tells us that Siriad is an old name for Egypt. He believes that the monument mentioned in Isaiah 19:19 refers to the Great Pyramid of Giza. The Great Pyramid is in the exact location between the ancient Upper Egypt and Lower Egypt. The natives who lived in the area of the Great Pyramid called it "Enoch's Pillar." Salt deposits were found in the queen's chamber, proving the pyramid itself has been under salt water, which makes it a pre-flood monument.[2]

As I said in speaking of the origin of astronomy, Hipparchus is believed to be the earliest known astronomer to recognize and assess the precession of the equinoxes. But documents dating before Hipparchus, relating to Aristarchus of Samos, show that he used precession to correlate the position of stars at different times or epochs. But was Aristarchus really the first? No. This science has shown up in a "document" dated back to 2623 BC or earlier. The Great Pyramid of Giza is that "document." It contains proof that those who built it knew of precession. But before I disclose how

THE FUTURE IS IN THE STARS

the pyramid does this, I want to give you some more information on this mathematical wonder so you will understand the importance of it.

The Great Pyramid itself contains no pharaoh's body, no treasure chamber, and no treasures. The following information indicates that the pyramid was not built for a pharaoh but to transmit information to a future generation, as Josephus says.

Its base covers 13.6 acres. Only a solid stone mountain could endure the pyramid's immense weight, and indeed a flat, solid, granite mountain happens to be located just beneath the surface of the ground directly under the pyramid.

The Great Pyramid is built on perfectly level ground. Its level is so exact that it varies less than half an inch over the 13.6 acres. This is an amazing feat for ancient surveyors and would have been difficult, if not almost impossible to match, even today with our advanced technology.

It is built to face true north (and the builders did not have a compass. They had only the stars.)

The pyramid is located at the exact center of the earth's land mass in the Giza plateau of Egypt. That is, its east-west axis corresponds to the longest land parallel across the earth, passing through Africa, Asia, and America. Similarly, the longest land meridian on earth—through Asia, Africa, Europe, and Antarctica—also passes right through the pyramid. Since the earth has enough land area to provide three billion possible building sites for the pyramid, the odds of its having been built where it is are one *in three billion*.

Like twentieth-century bridge designs, the pyramid's cornerstones have balls and sockets built into them.

The outside surface stones are cut within 0.01 (1/100th) of an inch of perfectly straight and at nearly perfect right angles for all six sides. And they were placed together with an intentional gap of 0.02 of an inch. Modern technology cannot place such twenty-ton stones with greater accuracy than those in the pyramid.

Even more amazing is that the 0.02 of an inch gap was designed to allow space for glue to seal and hold the stones together. A white cement that connected the casing stones and made them watertight is still intact and stronger than the blocks that it joins.

We know from geometry that there is a universal relationship between the diameter of a circle and its circumference. Consider this: The height of the pyramid's apex would be 5,812.98 inches, and each side is 9,131 inches from corner to corner (in a straight line). If the circumference of the pyramid is divided by twice its height (the diameter of a circle is twice the radius), the result is 3.14159, which just happens to be pi. Incredibly, this calculation is accurate to six digits. Thus pi was designed into it 4,600 years ago (or more). Pi is demonstrated many times throughout the pyramid.

Other numbers are also repeated throughout. Each of the pyramid's four walls, when measured as a straight line, is 9,131 inches, for a total of 36,524 inches. At first glance, this number may not seem significant, but move the decimal point over and you get 365.24. Modern science has shown us that the exact length of the solar year is 365.24 days.

The average height of land above sea level (Miami being low and the Himalayas being high), as can be measured only by modern-day satellites and computers, happens to be 5,449 inches. That is the exact height of the pyramid.

Note: The difference between the stated height of the pyramid here (5,449 inches) and the height of the pyramid's apex above (5,812.98 inches) is because the pyramid does not have a "cap" bringing it to a "point."

All four sides of the pyramid are very slightly and evenly bowed in, or concave. This effect, which cannot be detected by looking at the pyramid from the ground, was discovered around 1940 by a pilot taking aerial photos to check certain measurements. As measured by today's laser instruments, all of these perfectly cut and intentionally bowed stone blocks duplicate exactly the curvature of the earth. The radius of this bow is equal to the radius of the earth. This radius of curvature is what Newton had long been seeking when working out his mathematical formulas. (He actually went to the pyramid to investigate.)

THE FUTURE IS IN THE STARS

There's so much more that I hesitate to take the time (and space) to relate it all. Instead, I'll relate the short version.

As Dr. Johnson said, the pyramid is actually a timeline. Using the Jewish sacred inch to equal one year, the pyramid undoubtedly yields several dates by means of geometry and other mathematics, without mistake. These dates are represented by intersections of passageways and chambers within the pyramid.

Those dates are as follows:

2623 BC: The year the pole star Thuban lined up precisely with the entrance passage into the pyramid (on the day of the vernal equinox, to the second). The pyramid's "door" at the entrance had a centralized opening allowing only about one-third of a degree field of view from the bottom of the 334-foot long passage.

This is where I regress to speak of the "document" (my term) that speaks of precession before the time of Aristarchus. In 2004 (the time I give for entering into the Age of Aquarius), the present pole star, Polaris, on the day of the vernal equinox (to the second) lined up precisely with that very same opening as Thuban had in 2623 BC. How's that for representing precession?

Note: Most pyramidologists believe 2623 BC to be the date the pyramid was completed, based on measurements giving this date at the entrance. I believe the pyramid could have been built as early as the time of Enoch, as Josephus seems to indicate, because of the superior knowledge they had. Their advanced knowledge of astronomy and mathematics would have allowed them to build the pyramid with the shaft pointing where differing pole stars would be at any point in time they so desired.

These are the major dates indicated in the pyramid:

- 1453 BC: The exodus (by intersecting passageways)
- 2 BC: Jesus's birth (by the "Grand gallery")
- AD 33: Jesus's crucifixion (by the "Grand gallery")
- 3996 BC: Creation (by geometric extension of pyramid geometry)

- AD 2004: Vernal equinox (Age of Aquarius) (by the "Great Step" at the entrance to the "King's chamber")

Note: When fractions of an inch are used, precise dates are found, not just years.

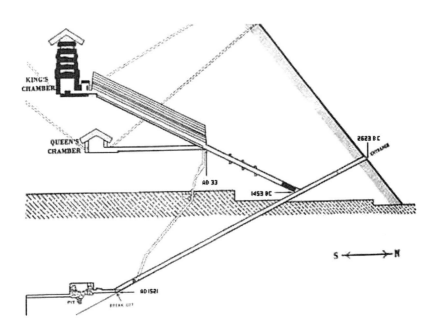

Credit: Russell, US Public domain.

At this point, I want to say that many who have investigated the pyramid have come up with other dates supposedly predicted by the pyramid. Since each mark at one-inch intervals represents a year in history, dates can be applied at any place along the line. Sheer coincidence dictates that a historical date would be indicated somewhere along that line at some point.

I compared those which correspond to historical events involving Israel against the zodiac *timeline* and would agree with those. Others, not involving Israel directly, I would consider fanciful fabrications of imagination.

I say this only to say that I believe the Great Pyramid was not built by Egyptians but by the antediluvian patriarchs to preserve the sciences of mathematics and astronomy, knowing that all would be lost in the flood, and to give us a timeline for the Age of Aquarius (the *age* in which something *big* would occur). Obviously, even they could not know "the day or the hour" of the second coming; therefore, they could not predict it.

Also, there are two main passages in the pyramid. If the angle of the ascending passage (26 degrees 18 minutes 9.63 seconds) above the horizon is applied to a map of the area so that a line is drawn 26 degrees 18 minutes 9.63 seconds above a horizontal (east-west) line that bisects the pyramid, that line then depicts the beginning and the end of the exodus of the Jews from where they were delivered from the Egyptians at the Red Sea to the crossing of the Jordan River into the Promised Land. This line also passes directly through the town of Bethlehem, the birthplace of Christ. As incredibly precise as this may seem, the pyramid actually pinpoints Christ's birthplace along this line.

So what is my point about the Great Pyramid? The antediluvian patriarchs had knowledge of astronomy and mathematics superior to ours and incorporated it into the pyramid building. The unmistakable dates indicated inside the pyramid agree with the dates I have put forward for the major historical events according to the zodiac. I believe, beyond any doubt, that their knowledge of the zodiac allowed them to do this. I believe that this is proof positive for a "biblical astronomy" timeline.

This is world history according to the zodiac!

1. Flavius Josephus. William Whiston, trans. *Antiquities of the Jews*. Project Gutenberg, 2009. http://www.gutenberg.org/. eBook.
2. Ken Johnson. *The Ancient Magic of the Pyramids*. Pocket Books, 1977.

CHAPTER 14

The Bottom Line

So then, what is the bottom line? Just this: I believe that time is short-time, that is, until the end of the present age. If time on earth before the second coming is as short as the zodiac may indicate, it's past time for all of us as Christians to *get on the ball,* do as God has called us to do, and be witnesses to as many as we can to bring them to salvation in Christ before it's too late (for them). Jesus said, "Go and make disciples of all nations" (Matthew 28:19–20 NRSV).

Let's tell them the gospel story and show them the story by the stars.

Post Script

By the way, as we have seen, even star clusters—not just individual stars making up constellations—are part of the story. Even nebulae are a part (according to their name). So get involved, find an astronomy club (or start one), and see the story "up close and personal" for yourself. You don't need a big, expensive telescope; a pair of binoculars will suffice. It's a great way to be a witness for Jesus and have fun at the same time.

Clear skies, and may God bless you in your stargazing.

Bibliography

Enoch. Schodde, George H., Trans. *Book of Enoch.* W. F. Draper, 1882.

Humphreys, Russell. "The Earth's Magnetic Field Is Young." *Acts & Facts* 22 (8). Institute for Creation Research, 1993.

Johnson, Ken. *The Ancient Magic of the Pyramids.* Pocket Books, 1977.

Josephus, Flavius. William Whiston, trans. *Antiquities of the Jews.* Project Gutenberg, 2009. eBook.

CPSIA information can be obtained at www.ICGtesting.com
Printed in the USA
BVOW05s1937080415

395350BV00001B/50/P